THE BEST OF CRAFTSMAN HOMES

THE BEST OF CRAFTSMAN HOMES

Gustav Stickley

➔

Peregrine Smith, Inc.

SANTA BARBARA AND SALT LAKE CITY

1979

Library of Congress Cataloging in Publication Data

Stickley, Gustav, 1858-1942.
  The best of Craftsman homes.

  Includes plans from the author's Craftsman Homes (1909)
and More Craftsman Homes (1912)
  1. Stickley, Gustav, 1858-1942. 2. Architecture,
Domestic—United States—Designs and plans. 3. Arts and
crafts movement. I. Stickley, Gustav, 1858-1942.
Craftsman homes. II. Stickley, Gustav, 1858-1942. More
Craftsman homes. III. Title.
NA737.S65A4   1979   728.3   79-14082
ISBN 0-87905-058-6

Design by Richard A. Firmage

Manufactured in the United States of America

INTRODUCTION

The Arts and Crafts Movement is most easily visible in its architecture, the California Bungalow, or what is also called the Craftsman Bungalow. The important figures in the history of architecture during the Arts and Crafts period are Charles and Henry Greene, Irving Gill, and Bernard Maybeck, all of whom built their major houses in California. But the person most responsible for making Craftsman architecture available to families of moderate income—and hence popularizing it—was neither an architect nor a Californian, but a cabinet maker named Gustav Stickley.

Stickley reproduced several hundred plans in the pages of his magazine, *The Craftsman,* for what he termed Craftsman Homes. The first Craftsman Home design appeared in the May 1903 issue. After providing the plan and a brief description of his five bedroom, two-story, fieldstone house, Stickley added: "To all subscribers of *The Craftsman* any processes or details incident to the building, finishing, or decoration of 'The Craftsman House' will be willingly given, through the correspondence department of the Magazine, or more directly by private letter."

In 1903 a brilliant designer named Harvey Ellis came to work for Stickley. Although he is best known for his inlaid furniture designs, he contributed greatly to Stickley's interest in architecture. Stickley reproduced Ellis's first house design in the July 1903 issue of *The Craftsman.* The next month Ellis was given the lead article, "An Urban House: Number Three of the Craftsman Series." It was clear that Stickley had begun to emphasize architecture in *The Craftsman,* and that he would publish regularly detailed plans in the magazine for his Craftsman Homes.

Stickley formalized his interest in architecture in the January 1904 issue. Perhaps through Ellis's suggestion, Stickley announced the founding of a Homebuilders' Club, to be conducted under the auspices of *The Craftsman.* Shortly after, he hired a staff of architects for another new department, Home Designing and Building. Stickley also formally announced his plans to publish, during each month of 1904, the "design of a detached residence of which the cost should range between two and fifteen thousand dollars." Each set of plans would include a description of building materials and the cost of the project. Stickley continued publishing these plans until the end of his career; in fact, the last four pages of the final volume of *The Craftsman* carried house plans.

Stickley described his new concept in architecture in the back of his 1906 furniture catalog:

> *The Craftsman Houses* are the natural outcome of the growth of *The Craftsman* idea. We began by making furniture; metalwork, textiles and needlework followed almost as a matter of course, and lastly came The Craftsman House, designed upon the same principles that render individual the other products of *The Craftsman Workshops.* All are akin in sturdy honesty of construction, beauty of finish and straightforward simplicity in meeting as fully as possible the needs for which they exist. While the unpretentious beauty of Craftsman furniture, metals and textiles harmonizes with almost any simple surroundings, the need of houses designed as a whole and embodying the strong, direct structural principles upon which all our work is based, was felt more and more as the movement advanced.

INTRODUCTION

More specifically, the Craftsman House should have no needless mouldings or projections, roofs should be designed to minimize repair caused by rust and rot, and exterior surfaces, if they are wooden, should be treated with oil or stain, instead of paint. Interiors must be planned "so that the greatest amount of space and freedom is secured within a given area," so Stickley insisted on as few partitions as possible, and those planned with no wasted space. "One large living room, well planned and structurally interesting," Stickley wrote, "can be made the center of a home life impossible to a house where the same space is divided into a number of little box-like compartments called by courtesy rooms...."

Another feature of the Craftsman House is the attention paid to woodwork and its color scheme. Woodwork provided the foundation for interior color: natural wood with a delicate tone of gray or green or brown cast over it. Wall color should harmonize with the woodwork. Stickley pointed to the decorative interest provided by structural features, such as beams, panels, wainscot, as well as built-in window seats, bookcases, cupboards, sideboards, cabinets and lockers. The exteriors of these houses differed widely. The bungalow became the most popular design—particularly in the west—but most of Stickley's designs were not bungalows. Stickley's Craftsman Houses borrowed from Mission style, English Tudor, Prairie School, and so on. He was simply less interested in exterior design than in interior space and the furnishing of that space.

Over the years Stickley expanded his architectural interests: he transformed the Home Builders' Club into a national organization so that families who were interested in building their own homes could meet in local chapters; and he started the Craftsman Mail Service, which provided free to members of the Home Builders' Club complete plans for needlework, leather, metal—in fact, all of the arts and crafts related to home decoration. The Service would also answer questions about decorating and furnishing a home, as well as providing landscape gardening ideas. Stickley also began featuring articles by or about Louis Sullivan, Charles and Henry Greene, and Irving Gill, as well as other articles about home building, furnishing, and care.

Through his efforts, Stickley generated interest in architecture. *House Beautiful,* in 1909, referred to Craftsman Houses as the new art in America. Stickley was receiving letters from readers all over the country who were building their own homes according to Stickley's plans. A number of pamphlets on the California Bungalow were being printed, the most notable among them, Henry Wilson's *The Bungalow Book* (Los Angeles, 1908). In 1909, Stickley edited, published and distributed his own *Craftsman Homes.* According to an advertisement in *The Craftsman, Craftsman Homes* was a "book of helpful suggestions to prospective home builders and to those already possessed of a home who wish to increase its beauty by simple and astute means. It contains plans for houses costing from $500 to $1500 which, it is obvious from the detailed descriptions, represent the best ideas in American domestic architecture." (Notice the drop in price from his initial 1904 announcement.)

Craftsman Homes sold for two dollars, bound in full linen crash. In addition to the illustrated descriptions of a large number of houses, there were articles of practical instruction in the making of Craftsman furniture, refinishing woods, and interior

INTRODUCTION

decorating. *The New York Times* and the prestigious *International Studio* both favorably reviewed the book. *Craftsman Homes* begins with an article by Edward Carpenter titled "The Simplification of Life"; simplification was the basis for Stickley's Arts and Crafts philosophy, and provided the foundation for his building designs. While most of *Craftsman Homes* was concerned with plans for small cottages, city houses, farm houses, club houses, houses made of cement, plaster, logs and shingles, there were several articles on laying out gardens, equipping kitchens, and decorating rooms. In addition, Stickley reproduced a selection of his furniture, needlework, and metalwork. Finally, in two long articles, Stickley described how to build some of his Craftsman furniture, and how to finish the wood.

In 1910 Stickley started a business called The Craftsman Home Building Company. According to his July 1910 furniture catalog, Stickley had built a number of houses around the New York area. The demand for his houses grew. By 1912, the entire 20,000 copies of *Craftsman Homes* was sold. And so great was the demand for another book of house designs that, in 1912, Stickley published *More Craftsman Homes*.

The new book contained illustrated descriptions of Stickley's "more recent designs in Craftsman houses suited for building in concrete, in stone, in brick and in wood. Many of these houses have already been built and have been found most satisfactory by their owners." Though there are some plans that date from 1907, most of the homes in *More Craftsman Homes* date from the years 1909 to 1911. Stickley paid less attention in this second volume to furniture and accessories; he seemed more concerned in spreading the word about Craftsman architecture and in making it available in a more rustic form. So there are many more plans reproduced in this second volume of log cabins and log bungalows.

The articles contained in *The Best of Craftsman Homes* were selected from *Craftsman Homes* and *More Craftsman Homes*. The selections try to present the range of houses Stickley designed over the years. The book also contains a selection of articles which formed the philosophical and aesthetic base for Stickley's *Craftsman Homes*. Think of the book, then, as an historical look at an important phase of American architecture. Or, use it the way Stickley intended, as an inventory of small, comfortable, simply-designed houses.

Craftsman architecture was not part of some passing fancy; Stickley reproduced plans for his homes in *The Craftsman* for over thirteen years. If we are to believe his own figures, over twenty million dollars worth of homes were built along Craftsman lines in 1915 alone, from Alaska to the Fiji Islands. The Craftsman Bungalow is still the predominant architecture in many areas of Los Angeles. And they still stand, weathered and painted, run-down and restored, in small towns across the country. They attest to a need, resurfacing today, for care and practical detail, functional design, for an uncluttered simplicity, and most of all, for quality in our daily living.

Barry Sanders

TABLE OF CONTENTS

TABLE OF CONTENTS

*Articles are from *More Craftsman Homes* (1912)

"THE SIMPLIFICATION OF LIFE:" A CHAPTER FROM EDWARD CARPENTER'S BOOK CALLED "ENGLAND'S IDEAL"

WHEN we remember the sincere reformers of the world, do we not always recall most gladly the simple men amongst them, Savonarola rather than Tolstoi, Gorky rather than Goethe, and would it not be difficult to associate this memory of individual effort for public good with consciously elegant surroundings. Could we, for instance, picture Savonarola with a life handicapped, perhaps, by eager pursuit of sartorial eccentricities, with a bias for elaborate cuisine and insistence upon unearned opulence, or the earning of luxury at the sacrifice of other's lives or happiness? It does not somehow fit into the frame. In remembering those who have dedicated their lives to the benefit of their own lands, we inevitably picture them as men of simple ways, who have asked little and given much, who have freed their shoulders from the burdens of luxury, who have stripped off from their lives the tight inflexible bandages of unnecessary formalities, and who have thus been left free for those great essentials of honest existence, for courage, for unselfishness, for heroic purpose and, above all, for the clear vision which means the acceptance of that final good, honesty of purpose, without which there can be no real meaning in life.

Such right living and clear thinking cannot find abiding place except among those whose lives bring them back close to Nature's ways, those who are content to be clad simply and comfortably, to accept from life only just compensation for useful toil, who prefer to live much in the open, finding in the opportunity for labor the right to live; those who desire to rest from toil in homes built to meet their individual need of rest and peace and joy, homes which realize a personal standard of comfort and beauty; those who demand honesty in all expression from all friends, and who give in return sincerity and unselfishness, those who are fearless of sorrow, yet demand joy; those who rank work and rest as equal means of progress—in such lives only may we find the true regeneration for any nation, for only in such simplicity and sincerity can a nation develop a condition of permanent and properly equalized welfare.

By simplicity here is not meant any foolish whimsical eccentricity of dress or manner or architecture, colonized and made conspicuous by useless wealth, for eccentricity is but an expression of individual egotism and as such must inevitably be short-lived. And what our formal, artificial world of today needs is not more of this sort of eccentricity and egotism, but less; not more conscious posing for picturesque reform, but greater and quieter achievement along lines of fearless honesty; not less beauty, but infinitely more of a beauty that is real and lasting because it is born out of use and taste.

From generation to generation every nation has the privilege of nourishing men and women (but a few) who think and live thus sincerely and beautifully, and who so far as possible strive to impress upon their own generation the need of such sincerity and beauty in daily life. One of the rarest and most honest of these sincere personalities in modern life is Edward Carpenter, an Englishman who, though born to wealth and station, has stripped his life of superfluous social paraphernalia and stepped out of the clumsy burden of tradition, up (not

1

down) to the life of the simple, common people, earning his living and that of his family as a cobbler (and a good one, too) and living in a peaceful fashion in a home planned and largely constructed by himself. His life and his work are with the people. He knows their point of view, he writes for them, lectures for them, and though a leader in modern thought in England and a man of genius, he is one with his daily associates in purpose and general scheme of existence. In all his present writings the common man and his relation to civilization, is Mr. Carpenter's theme, and he deals with the great problems of sociology in plain practical terms and with a straightforward thought born of that surest knowledge possible, experience.

From the beginning of the endeavor of THE CRAFTSMAN to aid in the interests of better art, better work and a better and more reasonable way of living, the work of Edward Carpenter has been an inspiration and an ideal, born out of that sympathy of purpose which makes men of whatever nation brothers and comrades. We have from time to time in the magazine quoted from Mr. Carpenter's books at length, feeling that he was expressing our own ideal as no words of ours could, and particularly have we felt a oneness of purpose with him in his book called "England's Ideal," in which he publishes a chapter on the "Simplification of Life," which with its honesty, sincerity, its high courage and rare judgment should make clear the pathway for all of those among us who are honestly interested in readjusting life on a plane of greater usefulness and higher beauty. In this essay which we purpose here to quote at length, Mr. Carpenter begins by speaking of his own method of readjusting his life as follows:

"IF YOU do not want to be a vampire and a parasite upon others, the great question of practical life which everyone has to face, is how to carry it on with as little labor and effort as may be. No one wants to labor needlessly, and if you have to earn everything you spend, economy becomes a very personal question—not necessarily in the pinching sense, but merely as adaptation of means to the end. When I came some years ago to live with cottagers (earning say £50 to £60 a year) and share their life, I was surprised to find how little both in labor and expense their food cost them, who were doing far more work than I was, or indeed the generality of the people among whom I had been living. This led me to see that the somewhat luxurious mode of living I had been accustomed to was a mere waste, as far as adaptation to any useful end was concerned; and afterward I had decided that it had been a positive hindrance, for when I became habituated to a more simple life and diet, I found that a marked improvement took place in my powers both of mind and body.

"The difference arising from having a small piece of garden is very great, and makes one feel how important it is that every cottage should have a plot of ground attached. A rood of land (quarter acre) is sufficient to grow all potatoes and other vegetables and some fruit for the year's use, say for a family of five. Half an acre would be an ample allowance. Such a piece of land may easily be cultivated by anyone in the odd hours of regular work, and the saving is naturally large from not having to go to the shop for everything of this nature that is needed.

"Of course, the current mode of life is so greatly wasteful, and we have come to consider so many things as necessaries—whether in food, furniture, clothing or what not—which really bring us back next to no profit or pleasure compared with the labor spent upon them, that it is really difficult to know where the balance of true economy would stand if, so to speak, left to itself. All we can do is to take the existing mode of life in its simpler forms, somewhat as above, and work from that as a basis. For though the cottager's way of living, say in our rural districts or in the neighborhood of our large towns, is certainly superior to that of the well-to-do, that does not argue that it is not capable of improvement. * * * *

"NO DOUBT immense simplifications of our daily life are possible; but this does not seem to be a matter which has been much studied. Rather hitherto the tendency has been all the other way, and every additional ornament to the mantelpiece has been regarded as an acquisition and not as a nuisance; though one doesn't see any reason, in the nature of things, why it should be regarded as one more than the other. It cannot be too often remembered that every additional object in a house requires additional dusting, cleaning, repairing; and lucky you are if its requirements stop there. When you abandon a wholesome tile or stone floor for a Turkey carpet, you are setting out on a voyage of which you cannot see the end. The Turkey carpet makes the old furniture look uncomfortable, and calls for stuffed couches and armchairs; the couches and armchairs demand a walnut-wood table; the walnut-wood table requires polishing, and the polish bottles require shelves; the couches and armchairs have casters and springs, which give way and want mending; they have damask seats, which fade and must be covered; the chintz covers require washing, and when washed they call for antimacassars to keep them clean. The antimacassars require wool, and the wool requires knitting-needles, and the knitting-needles require a box, the box demands a side table to stand on and the side table involves more covers and casters—and so we go on. Meanwhile the carpet wears out and has to be supplemented by bits of drugget, or eked out with oilcloth, and beside the daily toil required to keep this mass of rubbish in order, we have every week or month, instead of the pleasant cleaning-day of old times, a terrible domestic convulsion and bouleversement of the household.

"It is said by those who have traveled in Arabia that the reason why there are so many religious enthusiasts in that country, is that in the extreme simplicity of the life and uniformity of the landscape there, heaven—in the form of the intense blue sky—seems close upon one. One may almost see God. But we moderns guard ourselves effectually against this danger. For beside the smoke pall which covers our towns, we raise in each household such a dust of trivialities that our attention is fairly absorbed, and if this screen subsides for a moment we are sure to have the daily paper up before our eyes so that if a chariot of fire were sent to fetch us, ten to one we should not see it.

"However, if this multiplying of the complexity of life is really grateful to some people, one cannot quarrel with them for pursuing it; and to many it appears to be so. When a sewing machine is introduced into a household the simple-minded husband thinks that, as it works ten times as quick as the hand, there will now be only a tenth part of the time spent by his wife and daughter in sewing that there was before. But he is ignorant of human nature. To his surprise he finds that there is no difference in the time. The difference is in the plaits and flounces—they put ten times as many on their dresses. Thus we see how little external reforms avail. If the desire for simplicity is not really present, no labor-saving appliances will make life simpler.

"As a rule all curtains, hangings, cloths and covers, which are not absolutely necessary, would be dispensed with. They all create dust and stiffness, and all entail trouble and recurring expense, and they all tempt the housekeeper to keep out the air and sunlight—two things of the last and most vital importance. I like a room which looks its best when the sun streams into it through wide open doors and windows. If the furnishing of it cannot stand this test—if it looks uncomfortable under the operation—you may be sure there is something unwholesome about it. As to the question of elegance or adornment, that may safely be left to itself. The studied effort to make interiors elegant has only ended—in what we see. After all, if things are in their places they will always look well. What, by common consent, is more graceful than a ship—the sails, the spars, the rigging, the lines of the hull? Yet go on board and you will scarcely find one thing placed there for the purpose of adornment. An imperious necessity rules everything; this rope could have no other place than it has, nor could be less thick or thicker than it is; and it is, in fact, this necessity which makes the ship beautiful. * * * *

"WITH regard to clothing, as with furniture and the other things, it can be much simplified if one only desires it so. Probably, however, most people do not desire it, and of course they are right in keeping to the complications. Who knows but what there is some influence at work for some ulterior purpose which we do not guess, in causing us to artificialize our lives to the extraordinary extent we do in modern times? Our ancestors wore woad, and it does not at first sight seem obvious why we should not do the same. Without, however, entering into the woad question, we may consider some ways in which clothing may be simplified without departing far from the existing standard. It seems to be generally admitted now that wool is the most suitable material as a rule. I find that a good woolen coat, such as is ordinarily worn, feels warmer when unlined than it does when a layer of silk or cotton is interposed between the woolen surface and the body. It is also lighter; thus in both ways the simplification is a gain. Another advantage is that it washes easier and better, and is at all times cleaner. No one who has had the curiosity to unpick the lining of a tailor-made coat that has been in wear a little time, will, I think, ever wish to have coats made on the same principle again. The rubbish he will find inside, the frettings and frayings of the cloth collected in little dirt-heaps up and down, the paddings of cotton wool, the odd lots of miscellaneous stuff used as backings, the quantity of canvas stiffening, the tags and paraphernalia connected with the pockets, bits of buckram inserted here and there to make the coat "sit" well—all these things will be a warning to him. * * * *

"And certainly, nowadays, many folk visibly are in their coffins. Only the head and hands are out, all the rest of the body clearly sickly with want of light and air, atrophied, stiff in the joints, strait-waistcoated, and partially mummied. Sometimes it seems to me that is the reason why, in our modern times, the curious intellect is so abnormally developed, the brain and the tongue waggle so, because these organs alone have a chance, the rest are shut out from heaven's light and air; the poor human heart grown feeble and weary in its isolation and imprisonment, the liver diseased and the lungs straitened down to mere sighs and conventional disconsolate sounds beneath their cerements.

"There are many other ways in which the details and labor of daily life may be advantageously reduced, which will occur to anyone who turns practical attention to the matter. For myself I confess to a great pleasure in witnessing the Economics of Life—and how seemingly nothing need be wasted; how the very stones that offend the spade in the garden become invaluable when foot-paths have to be laid out or drains to be made. Hats that are past wear get cut up into strips for nailing creepers on the wall; the upper leathers of old shoes are useful for the same purpose. The under garment that is too far gone for mending is used for patching another less decrepit of its kind, then it is torn up into strips for bandages or what not; and when it has served its time thus it descends to floor washing, and is scrubbed out of life—useful to the end. When my coat has worn itself into an affectionate intimacy with my body, when it has served for Sunday best, and for week days, and got weather-stained out in the fields with the sun and rain—then faithful, it does not part from me, but getting itself cut up into shreds and patches descends to form a hearthrug for my feet. After that, when worn through, it goes into the kennel and keeps my dog warm, and so after lapse of years, retiring to the manure-heaps and passing out on to the land, returns to me in the form of potatoes for my dinner; or being pastured by my sheep, reappears upon their backs as the material of new clothing. Thus it remains a friend to all time, grateful to me for not having despised and thrown it away when it first got behind the fashions. And seeing we have been faithful to each other, my coat and I, for one round or life-period, I do not see why we should not renew our intimacy—in other metamorphoses—or why we should ever quite lose touch of each other through the æons.

"In the above sketch my object has been not so much to put forward any theory of the conduct of daily life, or to maintain that one method of living is of itself superior to another, as to try and come at the facts connected with the subject. In the long run every household has to support itself; the benefits and accommodations it receives from society have to be covered by the labor it expends for society. This cannot be got over. The present effort of a large number of people to live on interest and dividends, and so in a variety of ways on the labor of others, is simply an effort to make water run up hill; it cannot last very long. The balance, then, between the labor that you may consume and the labor that you expend may be struck in many different ways, but it has to be struck; and I have been interested to bring together some materials for an easy solution of the problem."

A WORD ABOUT CRAFTSMAN ARCHITECTURE

ROM the beginning of my work as a craftsman my object has been to develop types of houses and house furnishings that are essentially cheerful, durable and appropriate for the kind of life I believe the intelligent American public desires. It comes to me every day of my life that a home spirit is being awakened amongst us, that as a nation we are beginning to realize how important it is to have homes of our own, homes that we like, that we have been instrumental in building, that we will want to have belong to our children. And, of course, this means that the homes must be honest and beautiful dwellings; they must be built to last; they must be so well planned that we want them to last, and yet they must be within our means. The delusion that a really beautiful home is within the reach of only the very rich is losing ground, as is its sister delusion that only by the slavish imitation of foreign models is æsthetic satisfaction to be achieved. People are also awakening to the fact that beauty in a building is not merely a matter of decoration, a something to be added at will, but is inherent in the lines and masses of the structure itself.

The point of view of the New England farmer, whose instructions to the architect were: "I'll build my house, and you fetch along your architecture and nail it on," is no longer typical. Today if you find a farmer who is thinking about building a home, the chances are that he and his family and the town builder spend a lot of evenings around the farm dining table, poring over plans and blue prints, and probably sketches which the farmer himself has made. There is no suggestion about an Italian villa or a French chateau, but the farmer is probably saying, "We want a large room to live in; we want an open fire in it because it looks cheerful and the children like it; we want a kitchen that my wife won't mind working in, and we want the house light and warm and pretty." This is a great change from the old days, and is in line with the theory on which Craftsman architecture is founded,—namely, a style of building suited to the lives of the people, having the best possible structural outline, the simplest form, materials that belong to the country in which the house is built and colors that please and cheer.

The Craftsman type of building is largely the result not of elaboration, but of elimination. The more I design, the more sure I am that elimination is the secret of beauty in architecture. By this I do not mean that I want to think scantily and work meagerly. Rather, I feel that one should plan richly and fully, and then begin to prune, to weed, to shear away everything that seems superfluous and superficial. Practically every house I build I find, both in structural outline and in the planning and the adjustment of the interior space, that I am simplifying, that I am doing away with something that was not needed; that I am using my spaces to better advantage. All of this means the expenditure of less money and the gain of more comfort and beauty.

It is only when we to an extent begin at the beginning of these things that we come to know how much that is superfluous we have added to life, and how fearful we have been to be straightforward and honest in any artistic expression. Why may we not build just the house we want, so that it belongs to our lives and expresses them? I have, all too slowly, begun to realize that it is right to build houses as people wish them, to cut away ornament, to subordinate tradition, and to put into the structure and into the interior finish the features that the occupants will find comfortable and convenient, and which almost inevitably result in beauty for them. It seems to me that every man should have the right to think out the plan for his house to suit himself, and then the architect should make this plan into a reasonable structure; that is, the outline should be well-proportioned and the different parts should be brought together so that the structural perfection will result in decorative beauty. If, added to this simple reasonable structure, the materials for the house are so far as possible those which may be found in the locality where the house is built, a beauty of fitness is gained at the very start. A house that is built of stone where stones are in the fields, of concrete where the soil is sandy, of brick where brick can be had reasonably, or of wood if the house is in a mountainous wooded region, will from the beginning belong to the landscape. And the result is not only harmony but economy. Why should the man who lives on a hillside bring brick from a long distance when the most interesting of modern dwellings, the log house, is at his hand? Or, if the brick could be had from the kiln a few miles away, why seek logs which are made expensive by the long freight haul from far-away mountains, and which would not seem in any way harmonious with the country where trees are scarce?

Once having settled upon the style of house which must suit the lie of the land and the happiness of the owner, the arrangement of floor spaces is next in significance. First of all, do away with any sense of elaboration and with the idea that a house must be a series of cells, room upon room, shut away from all others. Have a living room, the "great room" of the house that corresponds to the old "great hall" of ancient dwellings. This space is the opportunity for people to come together, to sit around the fireplace, for there must always be an open fire. It is the room where people read or study or work evenings, or play or dance, as the case may be,—the place where the elderly members of the family will have the greatest comfort and contentment, and where the children will store up memories that can never die. This great room must be well lighted, it will have groups of windows that furnish cheerful vistas in the daytime, and it must be so planned that seats or divans circle the fireplace and bring, by the very structure of the house, the family into intimate, happy relationship. It is wise, of course, that the entrance to this room from out of doors should be through an entry way or vestibule, in order that drafts may not be felt and to furnish coat room and opportunities for the putting aside of heavy wraps, umbrellas, etc. This should be borne in mind especially in cold climates where the whole comfort of the room may be sacrificed to a too abrupt connection with out of doors.

In the planning of this first floor and the adjustment of the spaces I have as few entrances and doorways as possible. They are expensive; they use up space, prevent a look of coziness and lessen the opportunities for building in of interesting fittings. It is also economical and picturesque to group the windows, and always the built-in fittings, the bookcases, the corner seats should be adjusted to the light from the windows as well as the fireplace. But here, as in the outside structure, I find the process of elimination must be always borne in mind. I do away with everything that does not contribute to comfort and beauty. This is a safe rule. The charm of the living room can be greatly enhanced by the alcove dining room, a greater sense of space is added and all the things that are put in the dining room to make it beautiful contribute to the pleasure of the people who are sitting in the living room. Also, the pleasure in the dining room is enhanced by glimpses of the living room, its spaces, its open fires, its grouped windows. This does away also with one partition; it furnishes opportunity for the interesting use of screens, or for the half-partition, on top of which may be placed lines of books or jars of ferns, not expensive ornaments for the house, and adding greatly to the beauty of color and to the homelike quality.

The question of built-in fittings is one that I feel is an essential part of the Craftsman idea in architecture. I have felt from the beginning of my work that a house should be live-in-able when it is finished. Why should one enter one's dwelling and find that it is a barren uninviting prisonlike spot, until it is loaded with furniture and the walls hidden under pictures and picture frames? I contend that when the builder leaves the house, it should be a place of good cheer, a place that holds its own welcome forever. This, of course, can only be accomplished by the building in of furnishings that are essentially structural features, and by the planning of the finishing of the walls and the woodwork so that they are a part of the inherent beauty of the home, and not mere backgrounds for endless unrelated decorations. In my own houses I study the color of the interior when I am designing the house. I plan the woodwork so that it embraces the built-in fittings, so that every bookcase or corner seat is a part of the development of the woodwork. In no other way can a house be made beautiful, or the architecture of the interior be complete and homelike.

You cannot make your house and your furnishings two separate schemes of attractiveness and expect an harmonious whole. The reason that this has been so much done in America is because people have not owned their homes. Usually their furniture alone belongs to them, and *that* they have tried to select so that it would be pleasant and well related. They have adjusted it to the houses that they have chanced to live in as well as they could, until they have grown to feel that a house is one thing and furnishings quite another. This is especially true in city apartments, where people expect to remain only a few years before they move on to another set of inconveniences. The furniture which in one house was adjusted to mahogany and green walls is later on adjusted to yellow oak and pink walls. And so families have gone from one set of torturing surroundings to another, until it seems a miracle that any sense of color and proportion in house furnishings should survive.

As for my own houses, I realize that they more or less demand the sort of furniture that I have been in the habit of planning for them. Not because I hold to one narrow outlook of beauty, but because I cannot but see that most of the imitation antiques as well as the types of modern furniture made purely for department store sales are not adjusted to simple practical artistic home surroundings. In the planning of my houses I have so eliminated the superfluous in structure, in floor plans, in interior fittings, that furniture which is not well planned or is overornamented must of necessity seem out of place.

"More Craftsman Homes," which is the second book of houses that I have published, stands for my own ideal of house building. In other words, it shows the extent to which I have been able in my own, perhaps small, way, to achieve

A WORD ABOUT CRAFTSMAN ARCHITECTURE

beauty in architecture through this process of elimination. It makes clear how I feel about houses which are built on economical principles, on good structural lines, always with the ideal of beauty, always insisting upon the utmost comfort and convenience. The edition (20,000) of the first book, "Craftsman Homes," which was published over two years ago, is now exhausted. And so great has been the demand for a book of Craftsman houses that we have found it necessary, in order to meet the response of the people who are interested in this kind of architecture, to get out within the last few months the book to which this little talk forms the introduction. This book in some respects is scarcely more than a catalogue. It is merely a straightforward presentation of my more recent designs in Craftsman houses suited for building in concrete, in stone, in brick and in wood. Many of these houses have already been built and have been found most satisfactory by their owners. Several of them have been built on Craftsman Farms, my own home place in New Jersey. I feel that every time a Craftsman house is built I verify in my own mind my ideal of architecture; that is, beauty through elimination.

There can be no doubt in my mind that a native type of architecture is growing up in America. I am not prepared to say to what extent the Craftsman idea has contributed to it, but I do know, from a very wide correspondence, that people all over the country are asking for houses in which they may be comfortable, houses which will be appropriate backgrounds for their own lives and right starting points for the lives of their children. It is my own wish, my own final ideal, that the Craftsman house may so far as possible meet this demand and be instrumental in helping to establish in America a higher ideal, not only of beautiful architecture, but of home life.

THE RELATION OF CRAFTSMAN ARCHITECTURE TO COUNTRY LIVING

IN THE development of Craftsman architecture I have had in mind especially the need of better dwellings for suburbs and country, and although I have also designed houses for city and town, most of the plans are intended for a rural environment. I believe that on the right use of the land depends much of our national welfare, and that therefore farm life should be made not only effective and profitable but also pleasant. I realize that the normal existence is one which includes an all-round development of the faculties, a wholesome proportion of manual and mental labor, opportunities for spiritual growth; and so I believe that a form of building which makes for simplicity of housekeeping and provides ample chance for outdoor working and living will help to increase the health, happiness and efficiency of the people.

My effort, therefore, has been directed toward something that will make country life more interesting. I see no reason why people should not build comfortable houses in the country, and make for themselves the kind of surroundings that will prove an incentive and inspiration, instead of the ugly buildings, tawdry furnishings and the many inconveniences which now make farm life so unattractive. Why should the advantages of our civilization be confined to the cities and towns? If they are to be of real value to the people at large, is it not imperative that they should be shared by the guardians of those natural resources from which the city draws its strength? And is it not an inadequate and one-sided sort of progress which gives to one set of workers those comforts and conveniences which modern science has devised, and leaves the others in conditions of discomfort and drudgery?

This lack of balance, I believe, can be adjusted to a great extent by the right kind of rural architecture. In the case of the farm, not only is it possible to plan the buildings and arrange the work on a basis of economy and convenience, but it is also possible to make the interior of the house so attractive that the housewife as well as the farmer will find it a place of daily pleasure and contentment. But it is essential for this that we simplify most of our present complicated ideals of cooking, ornament, apparel and furnishing; that we construct more convenient and comfortable homes; that we employ labor-saving devices for the house as well as for the barns and the fields. Especially is this needed for the woman who now turns in disgust from the overwork and isolation of the country to the city with its artificial amusements. By the use of labor-saving devices, by more scientific methods of housekeeping, by the simplifying of ways of living and thinking, what is now a heartbreaking drudgery can be made a source of joy and pride.

The house can be so planned that it will be a factor in the growth and happiness of the people. The large living room with its central fireplace will form the nucleus of home life, a place for rest and entertainment, for the gathering of the family and the planning of whatever industries are being developed on the farm. The piazza at the back will serve to connect house and garden and encourage outdoor living, and a dining porch will permit the joy of meals served in the open. The provision of a summer kitchen will bring fresh air and brightness to many tasks that would be wearisome indoors. The laundry tubs may be placed here instead of down cellar, thus saving time and steps and making labor less irksome. Here also can be done the cooking, preserving and canning of fruits and vegetables, the cleaning of milk cans, the preparing of food for the stock. In short, both house and housework can be so adjusted as to make labor a pleasure and the country home a center of interest.

And with the bettering of conditions in home and farm a new spirit will enter into our tasks. We shall readjust our attitude toward work. Instead of submitting to it as one of life's necessary evils, we shall welcome it with courage and with joy, as the thing through which we get our greatest development.

Not only does such a type of architecture as that which we advocate form an incentive and inspiration to country living, but it tends to promote a coöperative spirit. People are willing to coöperate if they can get more comfort into their lives, and keep better in touch with progress. And in the necessary development of rural life, problems of lighting, water supply, sewerage, farm machinery, motive power, etc., as well as of social and educational needs, will have to be solved by coöperation. Then with the increase of common material interests there will come a strengthening of spiritual ties. In place of the old feeling of rural isolation we shall find a quickening of the recreative and intellectual life of the people. Community spirit and community pride will become factors in the betterment of rural conditions, until every dweller of township, village, farm and open country will enjoy a share in the responsibilities and privileges of happy community life, and so contribute to the progress of the nation.

It was with this point of view that I started to organize Craftsman Farms, to demonstrate what could be done to bring interest, efficiency and beauty into country living. While we have a number of buildings there already, we expect later on to have a good many more, and from time to time we shall publish in THE CRAFTSMAN pictures, plans and descriptions of the houses, stables and shops we may build. The illustrations and plans of the Log House which we are showing on page 147 will give some idea of what we have accomplished in this direction, and the latch string is always out for anyone who is planning to build or who cares to see what we are doing.

G. S.

A CRAFTSMAN HOUSE FOUNDED ON THE CALIFORNIA MISSION STYLE

E have selected for presentation here what we consider the best of the houses designed in The Craftsman Workshops and published in THE CRAFTSMAN during the past five years. Brought together in this way into a closely related group, these designs serve to show the development of the Craftsman idea of home building, decoration and furnishing, and to make plain the fundamental principles which underlie the planning of every Craftsman house. These principles are simplicity, durability, fitness for the life that is to be lived in the house and harmony with its natural surroundings. Given these things, the beauty and comfort of the home environment develops as naturally as a flowering plant from the root.

As will be seen, these houses range from the simplest little cottages or bungalows costing only a few hundred dollars, up to large and expensive residences. But they are all Craftsman houses, nevertheless, and all are designed with regard to the kind of durability that will insure freedom from the necessity of frequent repairs; to the greatest economy of space and material, and to the securing of plenty of space and freedom in the interior of the house by doing away with unnecessary partitions and the avoidance of any kind of crowding. For interest, beauty, and the effect of home comfort and welcome, we depend upon the liberal use of wood finished in such a way that all its friendliness is revealed; upon warmth, richness, and variety in the color scheme of walls, rugs and draperies, and upon the charm of structural features such as chimneypieces, window-seats, staircases, fireside nooks, and built-in furnishings of all kinds, our object being to have each room so interesting in itself that it seems complete before a single piece of furniture is put into it.

This plain cement house has been selected for presentation at the head of the list chiefly because it was the first house designed in The Craftsman Workshops and was published in THE CRAFTSMAN for January, 1904, for the benefit of the newly formed Home Builders' Club. Therefore it serves to furnish us with a starting point from which we may judge whether or not any advance has since been made in the application of the Craftsman idea to the planning and furnishing of houses.

It was only natural that our first expression of this idea should take shape in a house which, without being exactly founded on the Mission architecture so much used in California, is nevertheless reminiscent of that style, this effect being given by the low broad proportions of the building and the use of shallow, round arches over the entrance and the two openings which give light and air to the recessed porch in front. The thick cement walls are left rough, a primitive treatment that produces a quality and texture difficult to obtain by any other method and to which time and weather lend additional interest. The roof, which is low pitched and has a fairly strong projection, is covered with unglazed red Spanish tile in the usual lap-roll pattern with ridge rolls and cresting. The house, as it stands, is a fair example of the way in which the problem of the exterior has been solved by the combination of three factors: simplicity of building materials, employment of constructive features as the only

Published in The Craftsman, January, 1904

A CRAFTSMAN HOUSE BUILT OF CEMENT OR CONCRETE AFTER THE CALIFORNIA MISSION STYLE, WITH LOW-PITCHED TILED ROOF, ROUND ARCHES AND STRAIGHT MASSIVE WALLS. THE DECORATIVE EFFECT DEPENDS ENTIRELY UPON COLOR, PROPORTIONS AND STRUCTURAL FEATURES.

decoration, and the recognition of the color element which is so necessary in bringing about the necessary harmony between the house and its surroundings. In this case the walls are

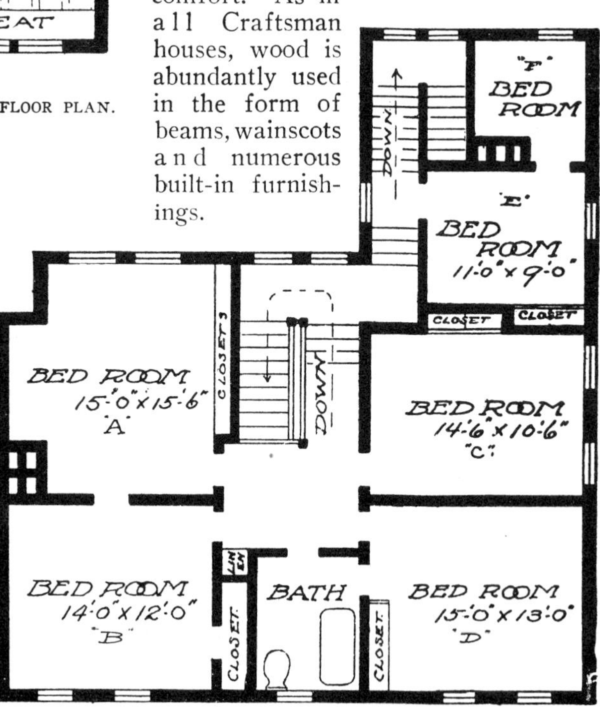

FIRST STORY FLOOR PLAN.

treated with a pigment that gives a soft warm creamy tone, almost a biscuit color, and the roof is dull red,—a scheme that is excellently suited to the prevailing color in California or in the South, where yellows, browns and violets abound. For the colder coloring of the northern or eastern landscape, the cement walls might either be left in the natural gray, or given a tone of dull green, which, applied unevenly, gives an admirable effect upon rough cast plaster. Or, for that matter, the house might be built of brick, stone, or of any one of the various forms of concrete construction. And the roof could be of tile, heavy shingles, or, if given a steeper pitch, of heavy, rough slate. In fact, the design as shown here is chiefly suggestive in its nature, making clear the fundamental principles of the Craftsman house and leaving room for such variation of detail as the owner may desire.

It will be noted that the foundation is not visible and that the turf and shrubbery around it appear to cling to the walls of the house,—a circumstance that is apparently slight and yet has a good deal to do with the linking of a house to the ground on which it stands. This effect would be greatly heightened by a growth of vines over the large plain wall spaces, which would lend themselves admirably to a natural drapery of ivy or ampelopsis.

The treatment of the interior is based upon the principles already laid down, the object being to obtain the maximum effect of beauty and comfort from materials which are few in number and comparatively inexpensive. Although we have not space here for illustration of the interior features, a description of the color scheme employed and of the use made of woodwork and built-in furnishings may serve to give some idea of its character. While the outside of the house is plain to severity, the inside, as we have designed it, glows with color and is rich in suggestion of home comfort. As in all Craftsman houses, wood is abundantly used in the form of beams, wainscots and numerous built-in furnishings.

SECOND STORY FLOOR PLAN.

A SMALL COTTAGE THAT IS COMFORTABLE, ATTRACTIVE AND INEXPENSIVE

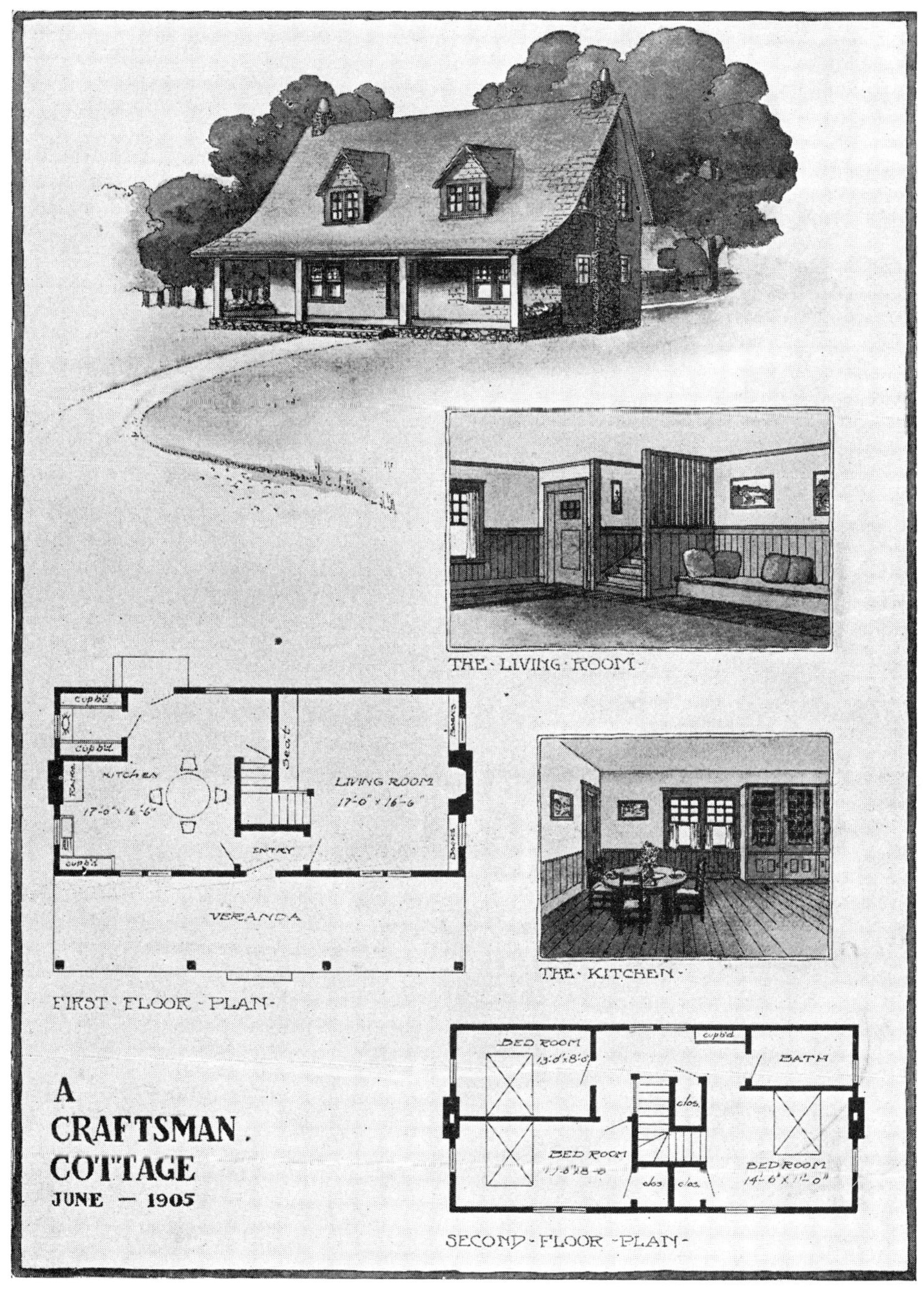

NOTE THE DIVISION OF SPACE SO THAT THE GREATEST AMOUNT OF FREEDOM AND CONVENIENCE IS OBTAINED WITHIN A SMALL AREA. THE ILLUSTRATIONS OF THE INTERIOR SERVE TO SHOW HOW THE STRUCTURAL FEATURES, ALTHOUGH SIMPLE AND INEXPENSIVE, GIVE TO EACH ROOM AN INDIVIDUAL BEAUTY AND CHARM. THE KITCHEN IS ARRANGED TO SERVE ALSO FOR A DINING ROOM.

A PLAIN HOUSE THAT WILL LAST FOR GENERA-
TIONS AND NEED BUT FEW REPAIRS

EXTERIOR VIEW SHOWING STRUCTURAL USE OF TIMBERS ON UPPER STORY AND EFFECT OF BUNGALOW ROOF.

MOST of the Craftsman houses are designed for an environment which admits of plenty of ground or at least of a large garden around them, but this one,—while of course at its best in such surroundings,—would serve admirably for a dwelling to be built on an ordinary city lot large enough to accommodate a house thirty feet square. Seen from the exterior, the house shows a simplicity and thoroughness of construction which makes for the greatest durability and minimizes the necessity for repairs. Also the rooms on both floors are so arranged as to utilize to the best advantage every inch of space and to afford the greatest facility for communication; a plan that tends to lighten by many degrees the burden of housekeeping.

In looking over the plan of the interior, we would suggest one modification which is more in accord with the later Craftsman houses. It will be noticed that the doors leading from the hall into the living room and dining room are of the ordinary size. We have found the feeling of space and freedom throughout the rooms intended for the common life of the family so much more attractive than the shutting off of each room into a separate compartment, so to speak, that were we to revise this plan in the light of our later experience, we would widen these openings so that the partitions would either be taken out entirely or else be suggested merely by a panel and post extending only two or three feet from the wall and open at the top after the fashion of so many of the Craftsman interiors. This device serves to break the space pleasantly by the introduction of a structural feature which is always decorative and yet to leave unhampered the space which should be clear and open.

While we advocate the utmost economy of space and urge simplicity as to furnishing, we nevertheless make it a point to render impossible even a passing impression of barrenness or monotony. As we have said, this is partly a matter of woodwork, general color scheme and interesting structural features that make each room a beautiful thing in itself, independent of any furnishing. But also we realize the never ending charm of irregularity in arrangement, that is, of having the rooms so placed and nooks and corners so abundant that the whole cannot be taken in at one glance.

In this case the simple oblong of the living room is broken by the window seat on one

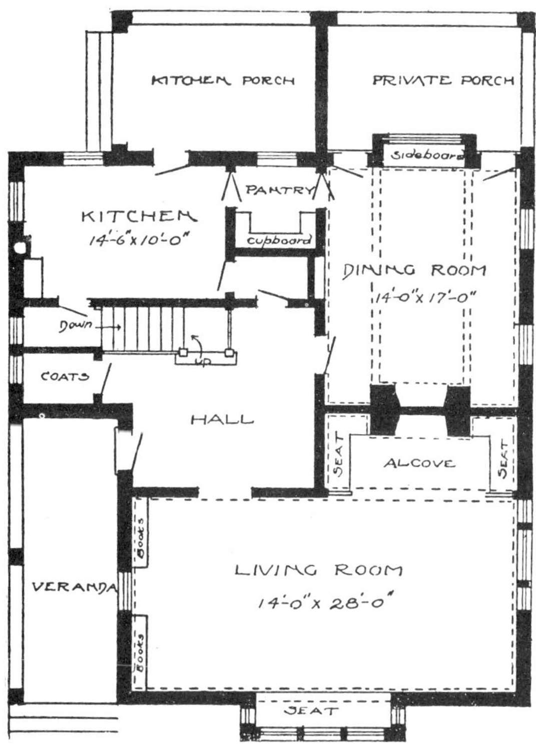

FIRST STORY FLOOR PLAN.

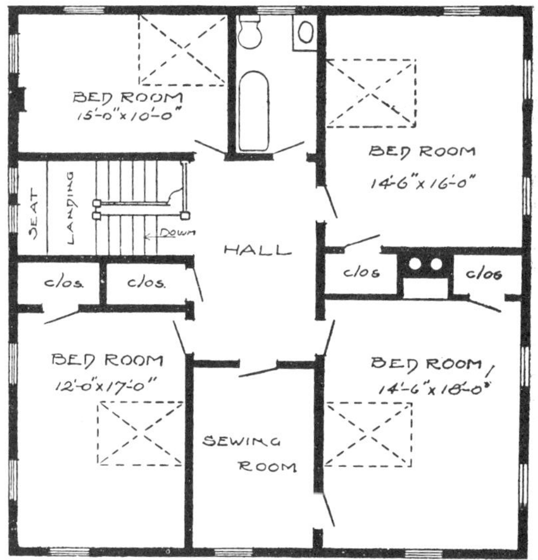

SECOND STORY FLOOR PLAN.

side and the alcove with its chimneypiece and fireside seats on the other. Just beside the alcove is a group of casement windows set high in the wall, so that the sill comes just on a level with the top of an upright piano. The same line is carried all around the room, which is wainscoted preferably with oak or chestnut.

END OF DINING ROOM, SHOWING EFFECT OF BUILT-IN SIDEBOARD, PICTURE WINDOW AND GLASS DOORS. A BUILT-IN CUPBOARD APPEARS AT THE SIDE OF THE ROOM.

FIRESIDE NOOK IN THE LIVING ROOM, SHOWING ARRANGEMENT OF SEATS AND THE PLACING OF A CRAFTSMAN PIANO JUST BELOW A GROUP OF CASEMENTS. THE DECORATIONS IN THE WALL PANELS ARE STENCILED ON ROUGH PLASTER IN COLORS THAT ARE MEANT TO ACCENT THE GENERAL COLOR SCHEME.

BEDROOM SHOWING A TYPICAL CRAFTSMAN SCHEME FOR DECORATING AND FURNISHING A SLEEPING ROOM. NOTE THE DIVISION OF WALL SPACES INTO PANELS BY STRIPS OF WOOD. THE PANELS ARE COVERED WITH JAPANESE GRASS-CLOTH.

A COTTAGE OF CEMENT OR STONE THAT IS CONVENIENTLY ARRANGED FOR A SMALL FAMILY

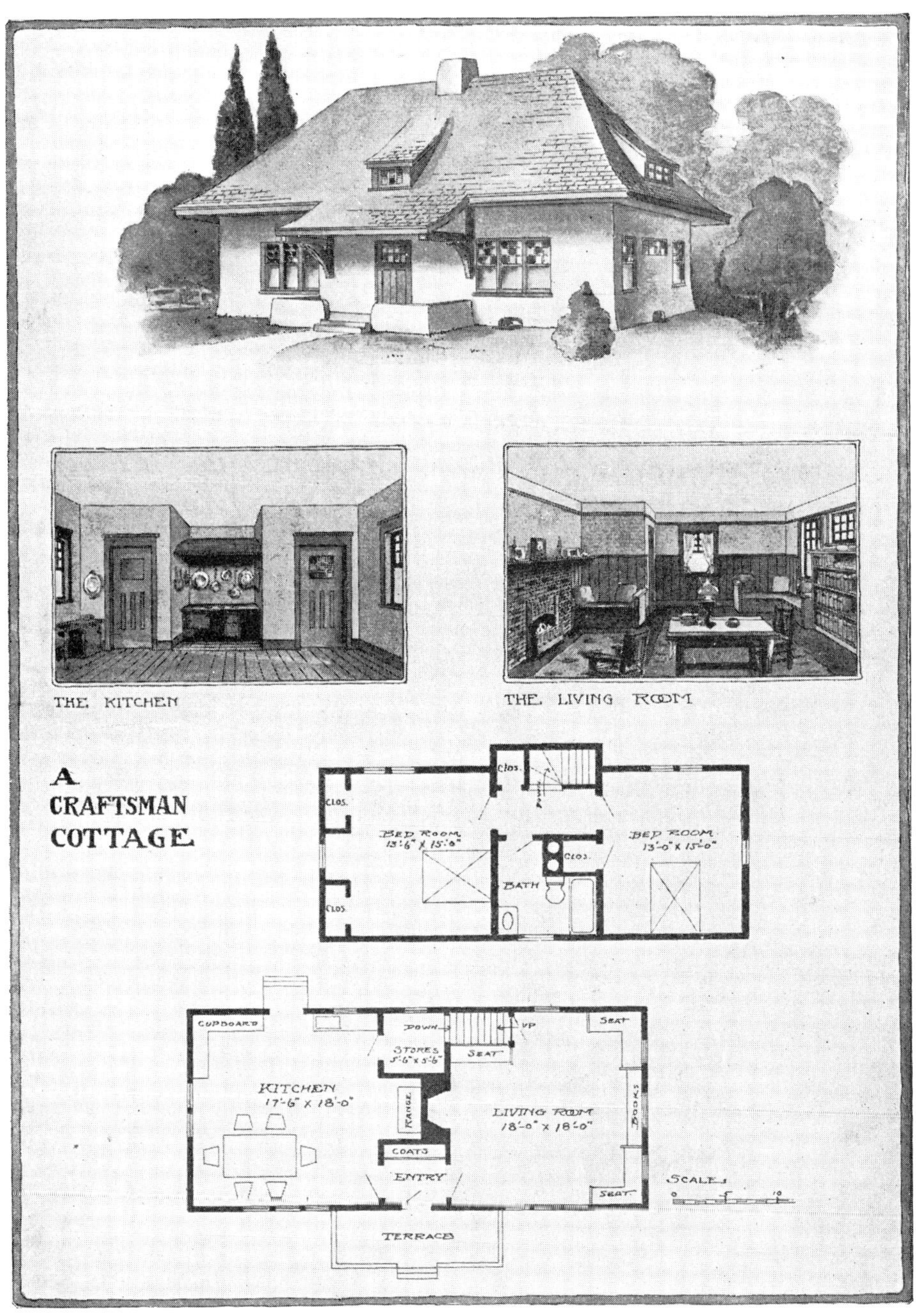

Published in The Craftsman, April, 1905.

THE DRAWING OF THE EXTERIOR SHOWS THE GRACEFUL LINES AND PROPORTIONS OF THE COTTAGE. THE GLIMPSES GIVEN OF THE INTERIOR SHOW HOW A HOODED RANGE IS PLACED IN A RECESS AND THEREFORE OUT OF THE WAY IN THE ROOM WHICH SERVES AS KITCHEN AND DINING ROOM, AND ALSO HOW A LIVING ROOM MAY BE MADE INDIVIDUAL AND CHARMING AT VERY LITTLE COST.

SUBURBAN HOUSE DESIGNED FOR A LOT HAVING WIDE FRONTAGE BUT LITTLE DEPTH

HOW THE HOUSE LOOKS WITH AMPLE GROUNDS AROUND IT AND A SETTING OF TREES FOR A BACKGROUND.

THIS house was designed primarily for use in the suburbs and the plan was adapted to a lot with wide frontage, but no great amount of depth. Of course, it would be better to have such a building surrounded by plenty of lawn, trees and shrubs; but if ground space were limited, a great deal could be made even of a meager allowance for front and back yards.

While the design admits the use of other materials which may be better suited to a given locality or considered more desirable by the owner, our plan was to have the house built of stone and shingles, the lower story and chimneys being of split field stone laid up in dark cement, and the upper story of cedar or rived cypress shingles, so finished that they are given a soft gray tone in harmony with the prevailing color of the stones. We have suggested that the shingle roof be stained or painted a soft moss green.

We regard the arrangement of these verandas as being especially comfortable and convenient, for although none of them are large, they serve admirably to supplement the inner rooms by furnishing what are practically outdoor rooms for general use. The front veranda, which is partially recessed, is sheltered from the street by the parapets and flower boxes. As doors open from this veranda into the hall, dining room and living room, it is much more closely connected with the house proper than is the case with the usual entrance porch, and is well fitted to serve as an outdoor sitting room. The veranda at the back of the house opens from the dining room and is meant to be used as a dining porch in summer time. Another door opening into the pantry makes it easy to serve meals out there. In winter this porch can easily be glassed in and used as a conservatory or sun room, and if heated, would make a very pleasant place for the serving of afternoon tea or for any such use. A third veranda opens from the kitchen and is meant especially for the comfort and convenience of the servants.

We would suggest here also that the openings from the hall into the dining room and living room be very much wider—a thing which could be easily done and which is now a feature of all the Craftsman houses. A glance at the floor plan will suggest the charm of such an arrangement, as it would allow a long vista from one fireplace to the other and would add much to the comfort and charm of the house as a whole. As will be noted, the living room fireplace is flanked on either side by a built-in bookcase with a casement window above, and in the dining room the same arrangement furnishes two china closets surmounted by casements set high in the

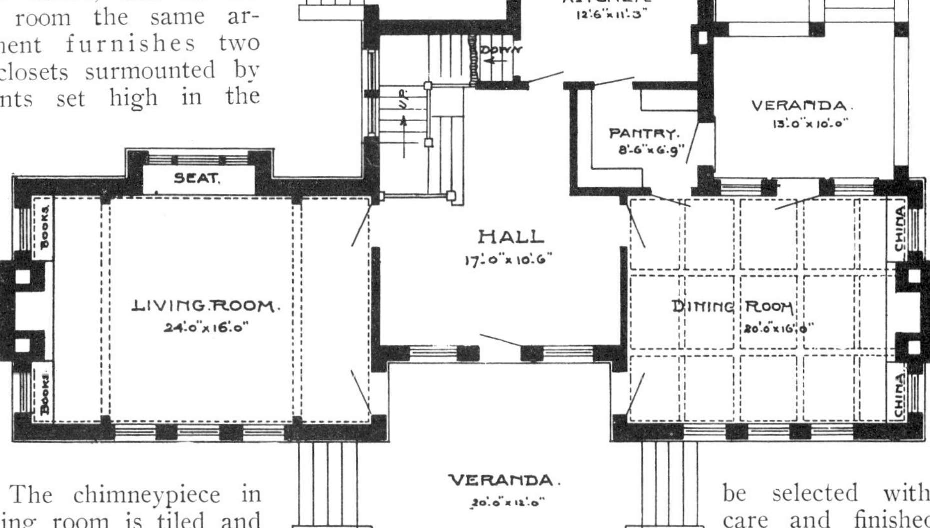

FIRST STORY FLOOR PLAN.

wall. The chimneypiece in the living room is tiled and the mantelpiece is on a level with the top of the wainscot, which runs around the room; but in the dining room the straight, massive brick chimneypiece runs to the ceiling, thus affording a pleasant variation in what otherwise might be too even a balance in the arrangement. The most decorative structural feature in the hall is the staircase, which is lighted by two casements set high above the lower landing and having wide sills, so that they afford an admirable place for plants.

The hall and dining room are wainscoted and the wall spaces in the living room are divided into panels by broad stiles of wood. As the woodwork is so essential in the decorative plan, it should be selected with great care and finished in a way to bring out all its charm of color, texture and grain. The general arrangement and style of the house would seem to demand some strong fibred, richly marked wood, which always seems best suited to rooms intended for general use.

The color scheme always is a matter of individual choice, but a safe rule to follow is to select some wood of rich and quiet coloring for the woodwork, and develop from that the color of the wall spaces, rugs and draperies.

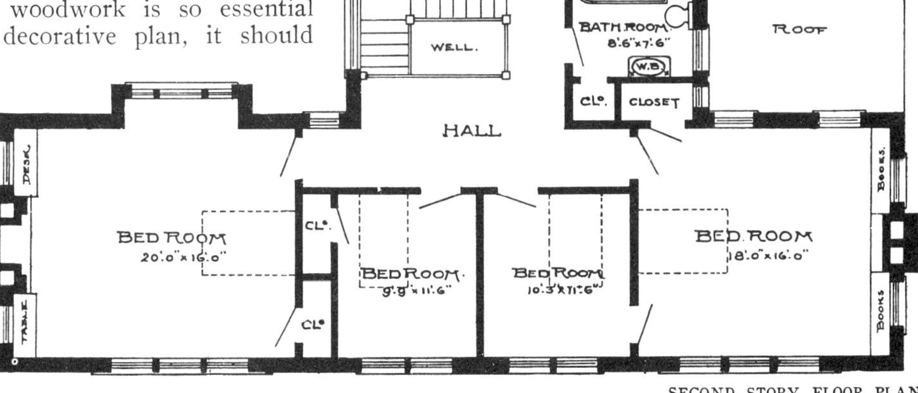

SECOND STORY FLOOR PLAN.

SUBURBAN HOUSE FOR WIDE LOT WITH LITTLE DEPTH

PARTIALLY RECESSED ENTRANCE PORCH, SO SHIELDED BY PARAPETS AND FLOWER BOXES THAT IT MAY BE USED AS AN OUTDOOR LIVING ROOM.

CORNER OF THE LIVING ROOM, SHOWING STRUCTURAL EFFECT OF FIREPLACE AND THE BUILT-IN BOOKCASES SET FLUSH ON EITHER SIDE SO THAT THE TOPS ARE PRACTICALLY AN EXTENSION OF THE MANTEL.

CEMENT HOUSE SHOWING CRAFTSMAN IDEA OF HALF-TIMBER CONSTRUCTION

EXTERIOR VIEW SHOWING STRUCTURAL USE OF TIMBERS AND THE WAY WINDOWS ARE BANDED TOGETHER.

A house that typifies to rather an unusual degree the Craftsman idea of construction is shown here. It is a perfect square in plan and is designed with the utmost simplicity. There are no bays, recesses or projections on the outside, the attractiveness of the exterior depending entirely upon the proportions of mass and spacing. It is a building which should attain the maximum of durability for cement con-

CEMENT HOUSE SHOWING HALF-TIMBER CONSTRUCTION

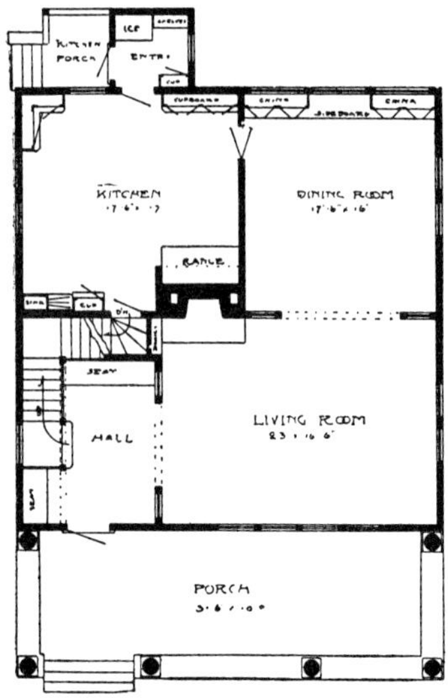

FIRST STORY FLOOR PLAN.

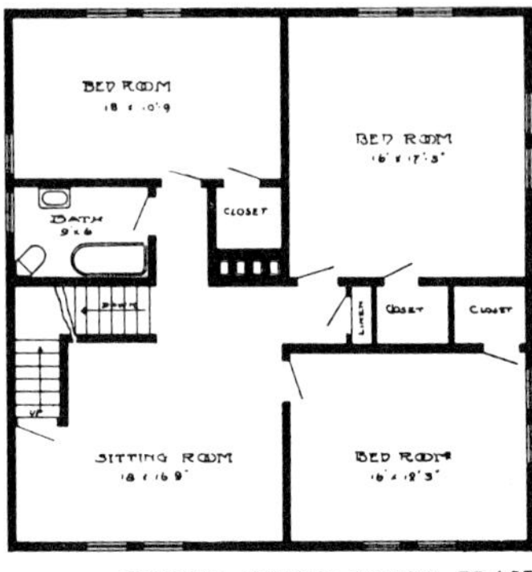

SECOND STORY FLOOR PLAN.

struction, as there is nothing to invite decay or render repairs necessary.

The walls are built of cement plaster and metal lath, the half-timber construction being used to break up the severely plain wall spaces into panels that are more agreeable to the eye. As originally designed the rough-finished cement was left in its natural gray color and the roof of white cedar shingles was merely oiled and left to weather to a harmonizing tone of silvery gray. The necessary color accent as well as the emphasis of form is given by the wood trim, which should be of cypress so treated that the brown color of the wood is fully brought out. The rafters of the porch as well as those supporting the widely overhanging roof are left uncased, carrying out the effect of solid construction which distinguishes the entire building and emphasizing the decorative use made of wood.

CORNER OF LIVING ROOM SHOWING FIREPLACE SET FLUSH WITH THE WALL AND HAVING PANEL OF DULL FINISHED PICTURE TILE. NOTE THE DECORATIVE EFFECT OF OPENINGS IN THE SPINDLE GRILLE WHICH APPEARS IN THE HALL, ALSO THE PLACING OF THE SEAT AND THE ARRANGEMENT OF THE STAIRCASE.

CEMENT HOUSE SHOWING INTERESTING ROOF TREATMENT AND ROOMY HOMELIKE INTERIOR

Published in The Craftsman, November, 1909.

CRAFTSMAN CEMENT HOUSE WITH EIGHT ROOMS AND THREE PORCHES: NO. 79.

L ONG, sloping roofs of shingle or slate, in which dormers are broken out to give the necessary height to the chambers, make the exterior of this cement house especially charming. The building is strongly constructed upon truss metal laths, and every care has been taken to avoid the possibility of leakage. The cement is brought close about the windows, which are so grouped as to break the wall into pleasing spaces.

The rooms are fitted with ample closets and are well lighted with large windows, both casement and double-hung. The amount of furniture that is built into the house will make quite a difference in the expense of furnishing it. In the kitchen there is a long dresser and a sink fitted with drip boards. A sideboard, flanked by china closets, is built into the dining room beneath the group of five small casements. The living room shows a long seat beneath the front windows, with built-in bookshelves on either side, and seats beside the piano in the opposite wall; but the most attractive feature is the deep inglenook which runs out between the twin porches that are connected with the room by means of

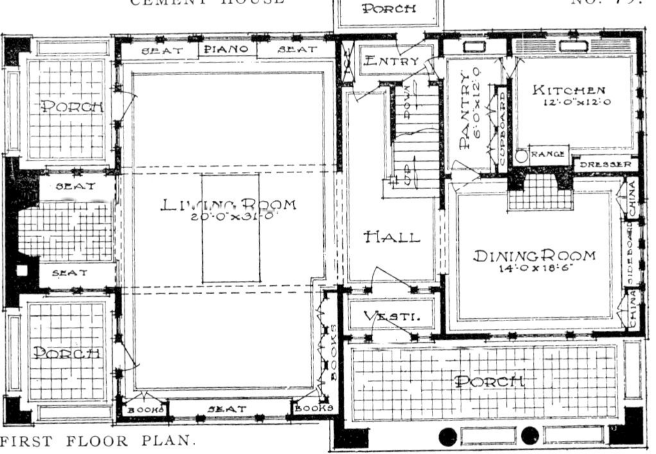

CEMENT HOUSE   NO. 79.

FIRST FLOOR PLAN.

25

long glass doors. The chimneypiece is of split field stone with rough tile hearth. On either side are two long settles with high wainscoted backs splayed out a little for greater comfort, and casement windows above. The ceiling within the nook is lower than that of the living room, being dropped to a level with the top of the heavy lintel across the entrance, adding a greater air of seclusion. From this lintel could be suspended Craftsman lanterns of hammered copper with amber glass. The nook, in fact, affords many opportunities for those little individual touches, in color, texture and detail, which can bring such comfort and beauty into an interior. And when the firelight plays over the warm tones of the woodwork and the soft varied colors of the stone, glinting upon a bit of metal above the mantelpiece, or upon the brighter colors of the cushions, the inglenook becomes truly the heart of the house, the center of interest of the home.

Turning to the floor plan of the upper story, we find that this is also worth studying, for it shows an unusually pleasant arrangement of the bedrooms. These are all light and airy,

and the windows, with their small panes, add greatly to the charm of the interior, and at the same time give a decorative touch to the outer walls. The most attractive feature of this plan is the built-in window seats which are seen in little nooks in four of the bedrooms, and which add so much to the comfort and cheerfulness of the rooms. The various corners, moreover, that break up the walls afford unique possibilities of furnishing.

VIEW OF THE FIREPLACE NOOK IN THE LIVING ROOM, SHOWING CHIMNEY-PIECE OF SPLIT FIELD STONE, ROUGH TILE HEARTH AND SETTLES WITH WAINSCOTED BACKS ON EITHER SIDE.

CRAFTSMAN CEMENT DWELLING INSPIRED BY OLD-FASHIONED NEW ENGLAND FARMHOUSE

Published in The Craftsman, September, 1909.

CRAFTSMAN CEMENT FARMHOUSE: NO. 74.

ALTHOUGH built according to a very modern method—cement on metal lath—this building will be seen to follow the salient structural features of the so-called New England farmhouse, varied, however, by the big dormer which breaks the long, sloping roof at the back to admit more light and air to the second story. The four-foot overhang at the eaves, with the deep brackets that support it, the large cement chimneys at either end, the porch at the corner and the pergola over the front door give interest to the exterior.

This pergola is better seen in the detail view of the building, on the next page. Instead of using the customary single heavy beam in the roof supports, we have shown here two smaller beams, thus making a lighter structure while taking nothing away from the strength of it. The pillars are of cement. The vine over the pergola and the flower-boxes set between the pillars add a note of grace and hospitality to the entrance, and seem to knit the house more closely to its surroundings. Vines could also be grown at the sides of the house, to soften the long straight lines of the two chimneys.

The entrance door, with its long metal hinges and knocker, and its row of small lights in the upper part of the panels, is as simple as it is decorative, and is quite in keeping with the rest of the construction. The casements on either side of the door serve to light the hall within, and at the same time add another note of welcome to the exterior.

The interior of the house, of course, meets the modern standards of comfort. The placing of the stairs, however, suggests the old New England arrangement; the landing is raised only a few steps above the living room and a railing runs along its edge so that the effect of a balcony is given. From the landing the stairs continue to the second story behind a partition of spindles, making them a part of both living room and hall, thus turning a necessary feature of the house into a most artistic one.

The large living room is well lighted by the window groups at the front and rear, and by the two large windows on each side of the

DETAIL VIEW OF CRAFTSMAN CEMENT FARMHOUSE NO. 74, SHOWING FRONT ENTRANCE WITH CEMENT PILLARS AND PERGOLA CONSTRUCTION ABOVE. THE SIMPLE BUT EFFECTIVE TREATMENT OF THE DOOR AND SMALL CASEMENTS ON EITHER SIDE IS ALSO WORTH NOTICING.

fireplace. A glass door leads onto the sheltered porch at the back, where meals may be served whenever the weather is warm enough. This porch also connects with the dining

HOUSE NO. 74: FIRST FLOOR PLAN.

room, which is provided with a built-in sideboard and communicates with the kitchen through a convenient pantry. The kitchen, it will be noticed, has ready access to the staircase, hall and entrance door. In fact the arrangement and relation of all the rooms has been made as simple and direct as possible in order to facilitate the work of housekeeping.

In one corner of the hall a coat closet is provided, and a wide opening at the right leads into a little den or library. In here is another open fireplace which utilizes the same chimney as the kitchen range, and on either side of the chimneypiece are built-in bookcases.

The upper floor plan gives one a sense of compactness and spaciousness combined. The rooms are large and airy, and amply supplied with closets. Two of the front bedrooms have built-in window seats—a feature which always adds much to the comfort and charm of an interior—and the bedroom to the right has the additional attraction of an open fire-

place which uses the same chimney as that of the den below. The large chamber on the left communicates with a sleeping porch at the rear which, being sheltered on three sides, is protected from the weather, and could be used practically the year round by those who believe in the healthfulness of outdoor sleeping. The large bedroom also communicates with the smaller central bedroom in front, which can also be entered from the hall. This little bedroom could be used as a dressing room, if desired, in connection with the adjoining room at the left. The bathroom is large and is provided with a linen closet.

All through the house, it will be noticed, in both the exterior and the interior, there is an entire absence of affectation or superfluous ornament. The treatment of the whole is extremely simple, and whatever decorative quality the building and the rooms possess will be found to be the outcome of necessary elements of the construction, handled in such a manner as to combine practical architectural features with

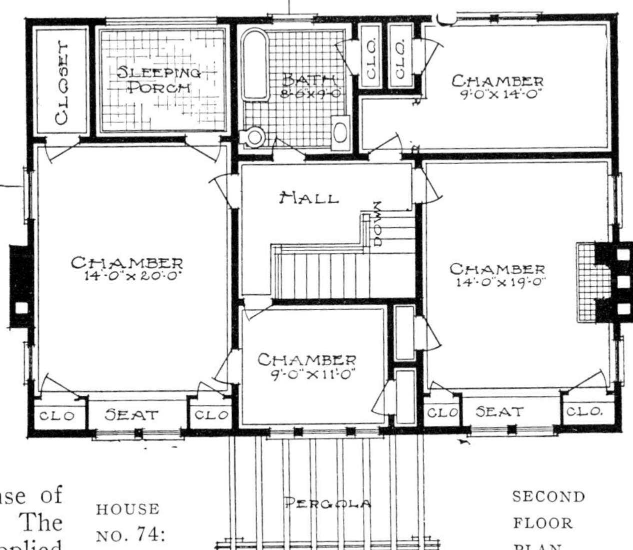

HOUSE NO. 74:

SECOND FLOOR PLAN.

comfort and usefulness of arrangement and harmony of proportion and line. Like all the houses shown here, the plans could be adapted to meet individual requirements.

SMALL ONE-STORY CEMENT BUNGALOW WITH SLATE ROOF, DESIGNED FOR A NARROW LOT

Published in The Craftsman, September, 1911.

CEMENT BUNGALOW FOR NARROW LOT: NO. 124

THIS one-story bungalow is a small, simply-arranged dwelling, intended for a narrow lot, and planned economically to afford the greatest possible comfort within a limited space. Cement plaster would be suitable for the walls and slate for the roof.

The most attractive feature of the exterior is the long porch which extends across the front of the house. This is enclosed by a low parapet of cement from which rise the hewn log pillars which support the pergola roof. Against this parapet is placed trelliswork, and from the trellis vines may be trained up the pillars and over the beams of the pergola above. Thus a very pleasant entrance will be formed, and even in winter, when the vines are leafless, the trelliswork and pergola construction overhead will still lend a distinctly decorative note to the exterior, softening by their graceful details the severity of what would otherwise be a plain cement building.

A touch of interest is also given to the side of the house by the extension of the main roof over the kitchen entrance, with its vine-encircled pillars rising from the cement floor.

Since the use of trelliswork is found to hold such charm, it might even be carried out further, and used in the entrance to the garden at the front of the house. The line drawing shown here gives a suggestion for such treatment, which would make a very effective gateway and would link the house more closely to its surroundings. Another factor in relieving the plainness of the bungalow walls is the use of small panes in all the windows. This not only breaks up the surface in a pleasing way, but at the same time seems to carry out the effect of the trelliswork.

The front door opens from the pergola porch directly into the ample living room, which is well lighted by the groups of windows on three sides, and is provided with an open fireplace. The latter, if a Craftsman fireplace-furnace, will serve to heat and ventilate the entire bungalow.

The kitchen and two bedrooms, though of moderate dimensions, will be large enough for a small family, and the whole plan is one that will lend itself to simple housekeeping.

The illustration of one of the rooms in this

30

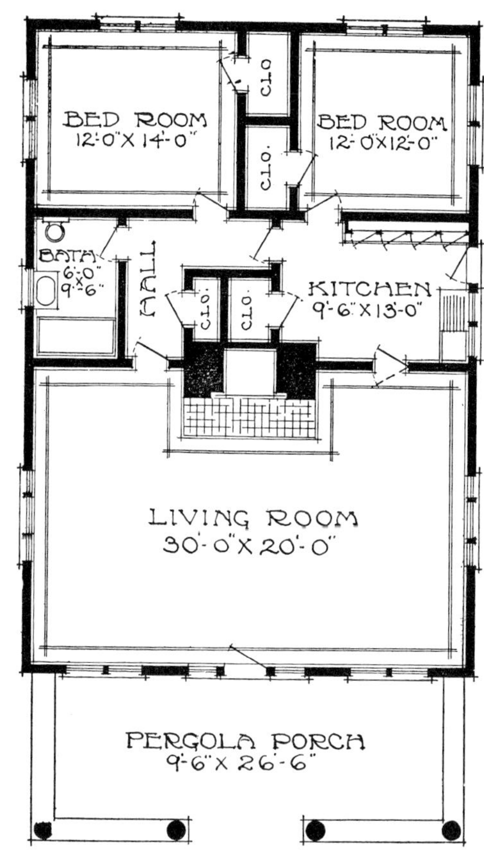

FLOOR PLAN OF BUNGALOW NO. 124.

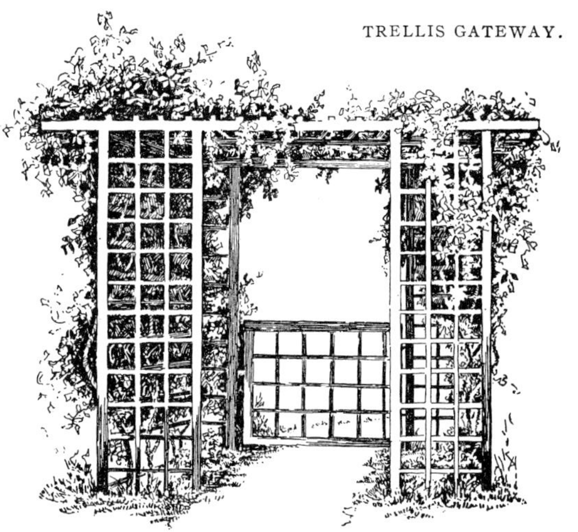

TRELLIS GATEWAY.

bungalow will give some impression of the effective simplicity of Craftsman bedroom furniture. The plain lines and restful proportions of the bed and bureau are thoroughly in keeping with the treatment of the woodwork, and the small square lights set in the paneled door form a pleasant relief to the flat surface of the walls. In the interior of this bungalow, as in other Craftsman houses, it will be found that much of the interest comes from the thoughtful handling of necessary architectural details, and the more carefully these are worked out, the easier will be the task of furnishing the home.

BEDROOM IN CEMENT BUNGALOW, NO. 124, WITH CRAFTSMAN FURNITURE.

INEXPENSIVE CEMENT AND SHINGLE COTTAGE

THIS is a simple little house intended for a small family, and the plan has purposely been arranged so that the construction shall be as inexpensive as is compatible with durability and safety. The shingled roof has a steep pitch and its line is broken by two shallow dormers on either side which afford plenty of light to the bedrooms and add to the interest of the exterior. The walls of the lower story are of cement on metal lath and the upper walls are shingled. A very satisfactory effect could be obtained by giving a rough, pebble-dash finish to the cement and brushing on enough pigment to give it a tone of dull grayish green, varied by the inequalities in the surface of the cement. It would pay to use rived cypress shingles for the upper walls, as these are much more interesting and durable than the ordinary sawn shingles, and possess a surface that responds admirably to the treatment with diluted sulphuric acid which we have found most successful with this wood. The roof could be either moss green or grayish brown, a little darker than the shingles of the upper walls. Four heavy cement pillars support the roof of the porch, which is also of shingles the same color as the main roof of the house. The porch floor may

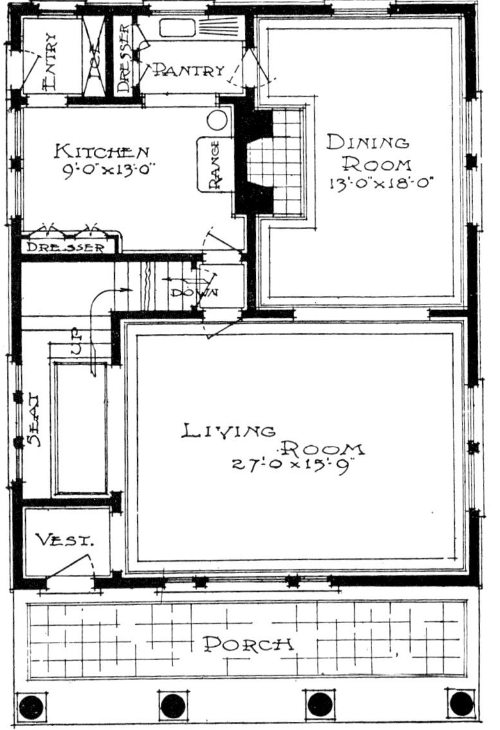

HOUSE NO. 84: FIRST FLOOR PLAN.

be paved with cement, or the outside edge might be plain cement of the same color as the walls, and the long strip down the center might be Welsh quarries or red cement marked off in squares.

The entrance door is at one end of the porch and opens into a small vestibule which leads directly into the living room, the opening being at right angles to the entrance door. A small partition separates the entrance from the stairway beyond, which is placed in a nook at the end of the living room. The entire end of this nook is occupied by a group of windows and a window seat. There is no fireplace in the living room, but in the dining room a large open fireplace uses the one central chimney which also serves for the kitchen range. The opening between the living room and the dining room is so broad that the fireplace serves equally well for both.

A small passageway leads from the kitchen to the living room, affording access to the entrance door, and in this passage is also the door leading to the cellar stairs. The kitchen is small and the pantry is little more than a nook in the larger room. Two large built-in china closets give plenty of room for the dishes, and the sink is placed in the pantry. Usually this arrangement would mean many additional steps, but the kitchen is so small that the distance from the range to the sink is no more than it would be in an ordinary room. An entry at the back of the kitchen communi-

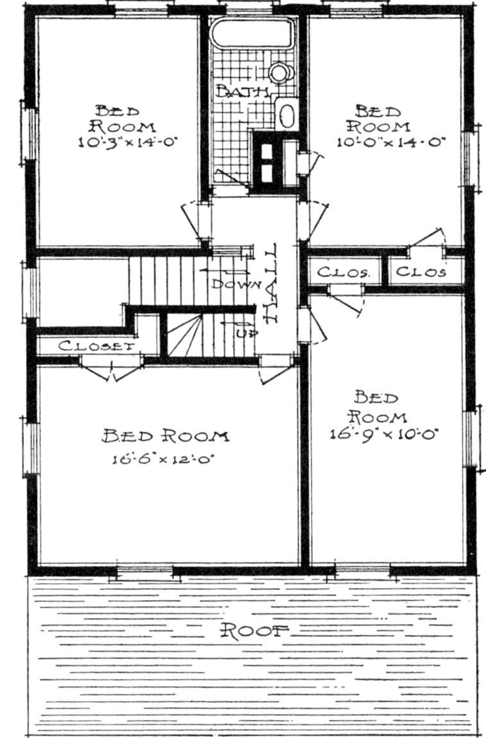

HOUSE NO. 84: SECOND FLOOR PLAN.

case in the center and the bathroom at the back.

On the whole, the interior of the cottage is one which would lend itself to simple housekeeping, and if tastefully furnished should prove both comfortable and attractive. The entrance could be made even more pleasant by the use of flower-boxes between the cement pillars of the porch, and vines planted along the side walls would add to the charm of the exterior and link the house more intimately to its surroundings.

The willow settle shown here would be a comfortable addition to this or any other interior, and would lighten up the general effect of the darker and heavier Craftsman oak furniture. The finish of soft green or deep golden brown would be a pleasant note in almost any color scheme, and the cushion coverings could be varied according to the material, color and design preferred.

LARGE WILLOW SETTLE WHICH HARMONIZES WELL WITH CRAFTSMAN OAK FURNITURE.

cates with a door leading to the outside, and furnishes a cool place for the ice box. A door from the pantry opens into the dining room.

The arrangement of the upper floor is very simple, as the four bedrooms occupy the four corners of the building with the hall and stair-

CRAFTSMAN HOUSE DESIGNED FOR NARROW LOT

CEMENT AND SHINGLE HOUSE WITH SIX ROOMS, PORCHES AND BALCONY FOR NARROW LOT: NO. 67.

THIS cement and shingle house, being only nineteen feet wide, can be built on the ordinary town lot. It is as compact and comfortable as possible for winter use, and still not without certain advantages in spring and summer which are quite lacking in the usual town block. Front and rear porches and a balcony that may be shaded by an awning will do much toward making the summer heat endurable.

The lower story of the house is of cement on a low foundation of split field stone, and the second story is covered with hand-split shingles.

The suggestion of pergola at the rear of the house is merely a three-foot projection on a porch running under the second story and is built of the exposed timbers of the house supported by pillars. It not only adds to the attractiveness of that corner as seen from the street, but, covered with vines, would give a lovely outlook for the dining-room windows, and, since a door connects it with the kitchen, may be used itself as a dining room in warm weather.

The stone chimney, instead of running up at an even depth from the foundation to the roof and narrowing above the fireplace on the ground floor, keeps its same width almost to the eaves, but slants in at the second story to about half the original depth. This does away with the monotonous line of the ordinary outside chimney and gives a fireplace upstairs as wide, although not so deep, as the one on the ground floor.

All the exposed windows on the second story are hooded to protect them from driving storms. It is an attractive feature in the construction, especially in connection with the window group—a long French casement flanked on either side by a double-hung window, looking out upon the balcony. The floor of this balcony and the timbers that support it form the ceiling of the porch. The ends of these exposed supports, projecting beyond the beam on which they rest, emphasize the line between the porch and the balcony and are at once decorative and economical, for the open construction does away with much repairing of the sort occasioned by the action of dampness upon timbers sheathed in.

The entrance door, which is very simple in design, opens from the front porch directly into the large living hall. Here, on each side of the open fireplace, we find built-in bookcases, with small casements set in the wall above. The room is also made light and cheerful by the window groups in the opposite wall and on each side of the entrance door, as well as by the wide opening into the dining room.

The view of the interior is made from a point just in front of the living room hearth and shows the use of spindles between the rooms and in the high balustrade that screens the two or three steps which lead up from the dining room, and are intended for the use of the servants. The meeting of these stairs with those from the living room makes an odd little corner that offers many possibilities for decorative effects. The dining room is wainscoted to the plate rail. The sideboard is built in and suggests the old-time dresser with its platter rail and side cupboards. There is a small pantry between the dining room and the kitchen, and the latter is fitted with the usual conveniences.

The upper floor plan shows a simple arrangement of the three bedrooms and bathroom.

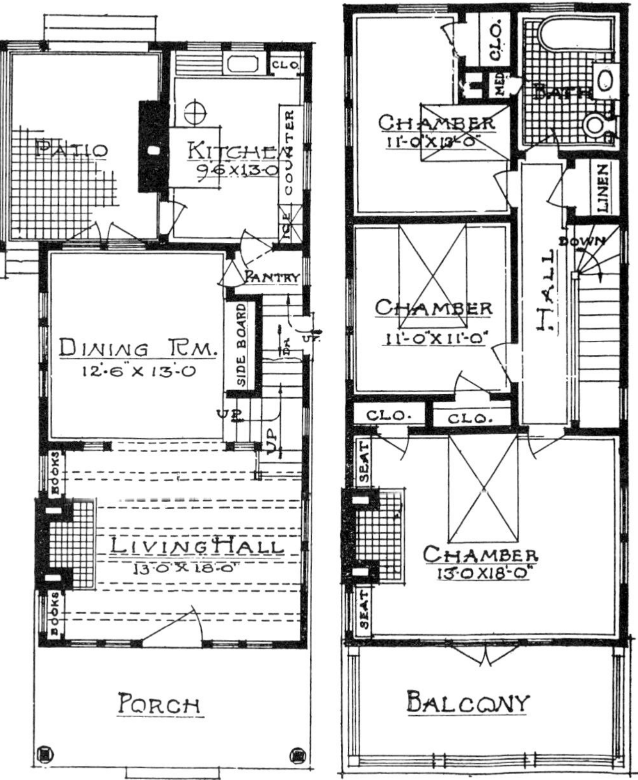

HOUSE NO. 67: FIRST AND SECOND FLOOR PLANS.

INTERIOR OF HOUSE NO. 67, SHOWING ONE CORNER OF THE LIVING HALL WITH GLIMPSE OF STAIRS AND DINING ROOM. THE DECORATIVE HANDLING OF WOODWORK AND STRUCTURAL FEATURES IS WORTH STUDYING.

SMALL BUT ROOMY ONE-STORY CEMENT BUNGA-LOW PLANNED FOR SIMPLIFIED HOUSEKEEPING

Published in The Craftsman, May, 1910.

ONE-STORY CEMENT BUNGALOW: NO. 90.

THIS one-story bungalow is meant for a small family, as it has room for only two bedrooms, but the arrangement of the interior is so compact that the maximum of room is afforded within the space enclosed by the outer walls. These are of cement on metal lath, with a roof of rough red slate and ridges of tile. The low, broad, sturdy effect is heightened by the use of buttresses which support the wide-eaved roof and give strength and dignity to the lines of the wall. The house has ample window space. Two small recessed porches at one end serve respectively as entrance porch and outdoor dining room. A glass door leads from the entrance porch directly into the living room.

The whole front of this room is taken up with the central group of windows and the casements set high on either side. A window seat is built below the middle group and bookcases occupy the remainder of the wall space to the height of the casements, and open shelves are built in on either side of the fireplace. The dining room, as is nearly always the case in a Craftsman house, is really a recess in the living room. A sideboard occupies the whole of the outside wall, with three casement windows set high above it. A glass door leads to the front porch, and the whole of the rear wall is taken up by casement windows and another glass

DETAIL OF CRAFTSMAN RECLINING CHAIR WITH ADJUSTABLE BACK AND SPRING CUSHION SEAT, SHOWN ON THE NEXT PAGE IN INTERIOR VIEW OF BUNGALOW.

SMALL BUNGALOW PLANNED FOR SIMPLE HOUSEKEEPING

CORNER OF LIVING ROOM IN ONE-STORY BUNGALOW NO. 90, SHOWING OPEN SHELVES ON EACH SIDE OF THE FIREPLACE, AND EFFECTIVE USE OF CRAFTSMAN FURNISHINGS.

door leading to the rear porch. The room is thus well lighted and cheerful.

A tiny hall opening from the other end of the living room gives access to the two bedrooms and also to the kitchen, which by this means is entirely shut off from the remainder of the house. The bath is so placed that it is accessible from both bedrooms and from the kitchen.

The arrangement of this cottage is such that the house-mistress is practically independent of servants, for the compactness of the floor plan, the directness of the communication between the several rooms, and the convenient placing of the tables, closets, dressers, etc., in the kitchen and pantry, all help to minimize the necessary household work.

Another important factor is the presence of the several built-in pieces — sideboard, seat and bookshelves. These not only reduce the amount of furniture needed for the living room and dining room, but add considerably to their comfort and charm.

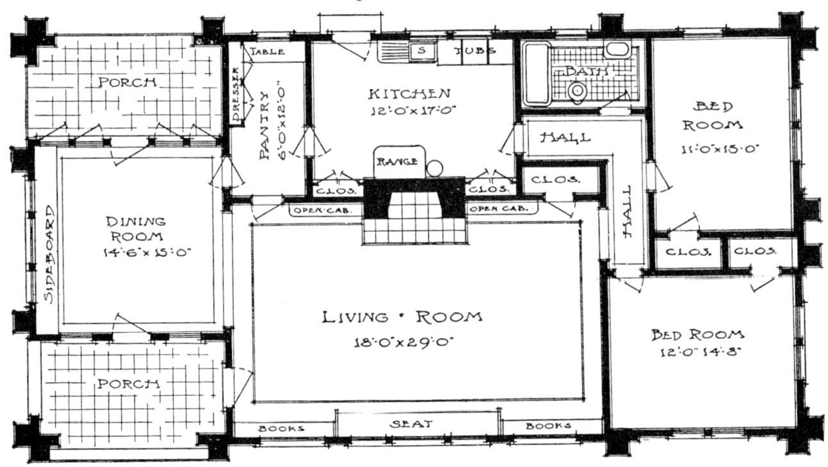

FLOOR PLAN OF ONE-STORY BUNGALOW: NO. 90.

TWO-STORY CEMENT BUNGALOW WITH AMPLE PORCH ROOM AND COMFORTABLE INTERIOR

Published in The Craftsman, December, 1909.

TWO-STORY CEMENT BUNGALOW WITH FIVE ROOMS AND PORCH: NO. 81.

WHILE this cottage is more elaborate in design than the one-story bungalow shown on the preceding page, the same general construction is used, cement mortar upon metal lath being chosen for the walls, stone for the chimney, logs for the pillars of the porch and cinder concrete for the porch floor. In this cottage, however, the roof is shingled. The exterior, though quite unpretentious, is pleasing in line and proportion, with its well-placed windows, sloping dormer and sheltered angle of the long, roomy porch.

The illustration of the living room will also furnish an idea of the appearance of the living room of the previous bungalow, inasmuch as the main structural beams are the same. Indeed, all the woodwork in the living room, with the exception of the baseboard, which is cut in between the studs, is simply the necessary structural beams. The stairs lead up from the right, and a curtain may be hung to shield those about the hearth from any draught that may come from the upper rooms.

A door from one corner of the living room

VIEW OF STONE FIREPLACE IN LIVING ROOM OF TWO-STORY BUNGALOW, NO. 81. THE WOODWORK SHOWN IN THIS INTERIOR IS SIMPLY THE STRUCTURAL BEAMS OF THE BUILDING.

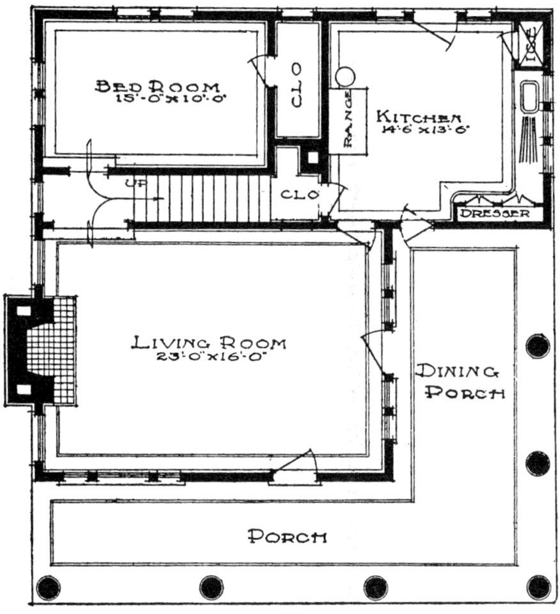

CEMENT BUNGALOW NO. 81: FIRST FLOOR PLAN.

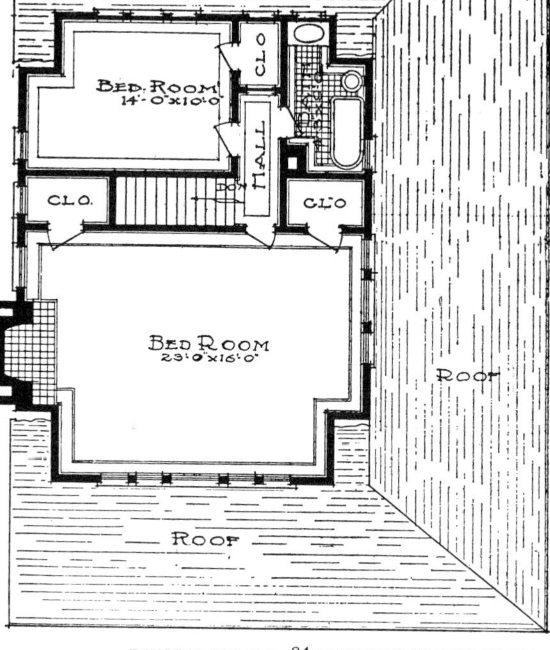

BUNGALOW NO. 81: SECOND FLOOR PLAN.

leads into the kitchen, which is large, well lighted and conveniently arranged. On the first floor there is also a bedroom provided with a large closet. The smaller closet beside it opens into the kitchen.

On the upper floor are two bedrooms and a bath, leading out of a small hall. The front bedroom is a large, cheerful apartment, with plenty of windows, and provided with an open fireplace which uses the same chimney as the one in the living room just below. Two good-sized closets are placed at opposite corners. Both this bedroom and the one in the rear could be made charming and comfortable by the use of window or corner seats and simple furnishings.

The arrangements of the rooms on each floor are so simple and compact that it would be an easy matter to keep them in order, and the accommodation would be quite sufficient for a small family.

Both this bungalow and the one previously described are intended exclusively for summer use. Either of them, however, could be built with an inside wall which would fit them also for winter, but this, of course, would add greatly to the expense. One of the chief advantages of both constructions shown is that when closed for the winter there is no place in which mice would build their nests or mildew collect. Every part of the cottage is open to the air. Returning in the spring, the owner needs only to brush down the cobwebs and wipe away the dust to find himself settled and at home for the summer.

CANDLESTICK, ANDIRONS AND BOWL FOR A CRAFTSMAN INTERIOR.

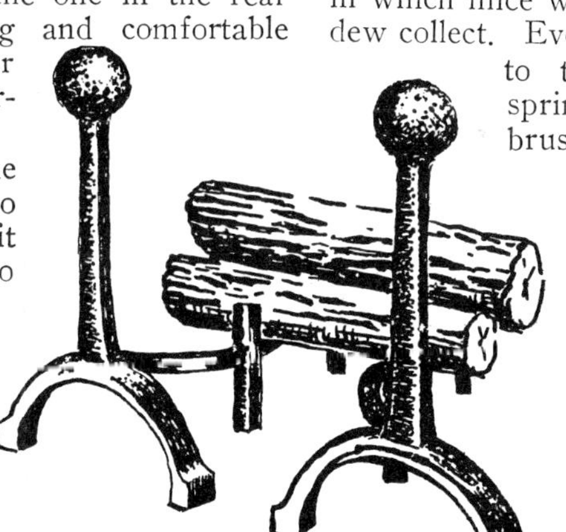

THESE FITTINGS ARE SO SIMPLE AND ARTISTIC THAT THEY WILL HARMONIZE WITH ANY FURNISHINGS.

CRAFTSMAN HOUSE DESIGNED FOR CITY OR SUBURBAN LOT

CEMENT stucco on metal lath is used for this house, which is intended for a city or suburban lot. The wide pergola and balcony, the stone and brick chimneys and groups of casements present a homelike and pleasing appearance. Purposely we have set the house down so as to show only a suggestion of the foundation. Instead of having a cellar, we prefer to level up the space between the walls of the foundation with earth, topping this with cinders and cinder concrete to a level of the foundation walls, and using 2 x 4's embedded in this for the first floor beams.

The interior has been arranged to eliminate all unnecessary partitions, and the stairs lead up directly from the living room. A den or workroom has been provided off the living room. Seats are built in beside each fireplace, and the one in the dining room has been so placed as to serve in connection with the table. Placing the table in this position allows ample space around the fireplace and does not give an appearance of being crowded, so often the case when fireplaces are built in the dining room. The house may be heated and ventilated by using Craftsman fireplace-furnaces.

The view of the fireplace end of the living room, with its corner seat, built-in bookshelves, and casement windows above, gives some idea of the general effect of the interior. The treatment of woodwork and wall spaces is simple but effective, and serves as a restful background for the various furnishings. The tiled hearth, the decorative placing of the bricks in the chimneypiece, and the recess above the shelf, add to the attraction of the inglenook, which, with its comfortable seats and inviting books, naturally becomes the center of interest of the room. The walls could be left plain or could be stenciled as suggested in the illustration with some design that would help to carry out the general color scheme of the interior. The woodwork of the staircase in the opposite corner of the living room could also be handled in such a way as to give structural beauty to this necessary feature.

COMPACTLY PLANNED, EIGHT-ROOM CEMENT HOUSE

CORNER OF LIVING ROOM IN CRAFTSMAN CEMENT HOUSE NO. 117, SHOWING THE FIREPLACE-FURNACE AND BUILT-IN SEAT AND BOOKSHELVES IN THE INGLENOOK.

The second floor is quite simple in arrangement, a bedroom occupying each of the four corners of the plan, with the bathroom between the smaller bedrooms on one side. Three of the bedrooms open into the hall, which is wide and cheerful and terminates a few steps down on the stair landing at the rear. From this landing several steps lead up to the other back chamber. All four of the bedrooms have windows on two sides, and ample closet room is provided.

Perhaps one of the most attractive features of this upper floor plan is the little balcony in the front over the pergola porch. Part of this balcony is open, protected by a railing, and part is recessed, as shown.

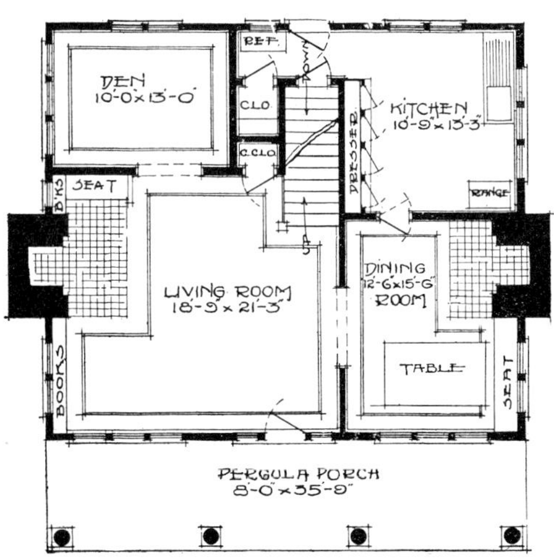

CEMENT HOUSE NO. 117: FIRST FLOOR PLAN.

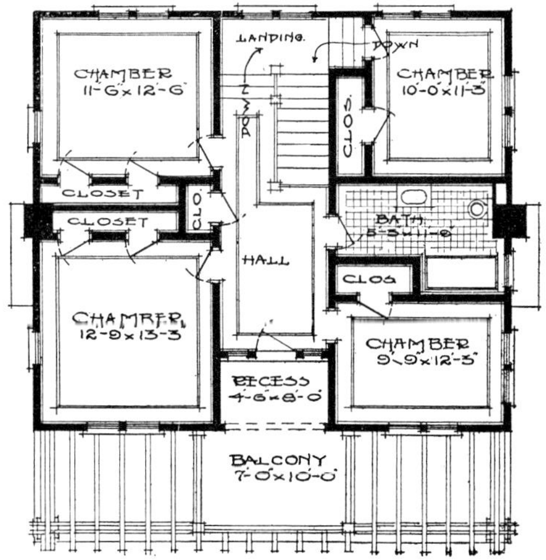

HOUSE NO. 117: SECOND FLOOR PLAN.

CEMENT COTTAGE FOR A NARROW TOWN LOT

Published in The Craftsman, June, 1911.

CRAFTSMAN CEMENT COTTAGE NO. 118.

THIS cement cottage is planned for a narrow lot and is only a story and a half high. It has a long roof line broken with flat dormers front and rear. The groups of windows are most interesting, all being casement except the large plate glass picture window of the front group, which is stationary. No front veranda has been provided; but the entry is recessed and the graceful arch emphasizes the cement construction. Hollyhocks would be especially charming against the plain walls.

On entering, you find the hall space has been included in the living room, with an open stair conveniently located near the entrance. A partition dividing dining and living room is only suggested—an arrangement which permits of a vista from living room through dining room and across the rear porch. Open bookshelves break up the long wall of the living room and a space has been left for the piano, which will give it an appearance of being built in.

The Craftsman fireplace-furnace is large and generous, and with the inviting seat nearby becomes at once the center of interest. The

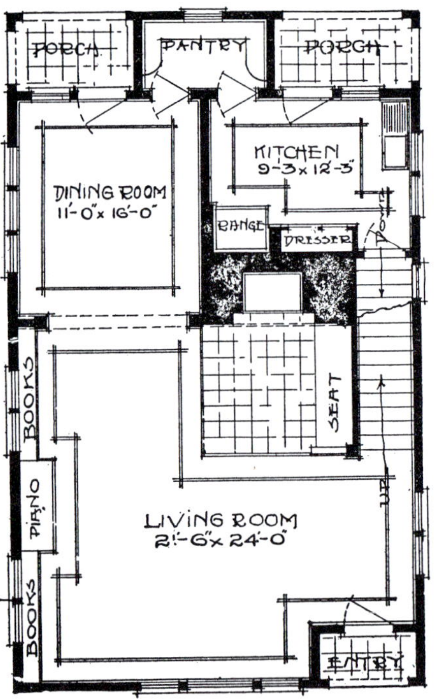

HOUSE NO. 118: FIRST FLOOR PLAN.

CEMENT COTTAGE FOR A NARROW TOWN LOT

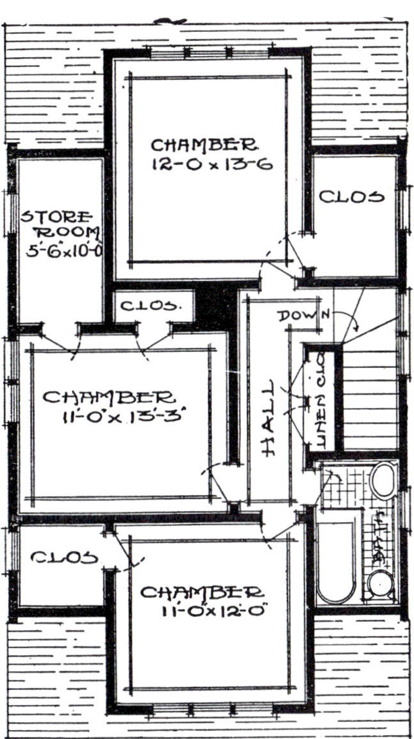

CRAFTSMAN CEMENT COTTAGE: SECOND FLOOR PLAN.

CRAFTSMAN LANTERN WHICH COULD BE USED IN THE INTERIOR SHOWN BELOW.

fireplace is built of common brick and is plastered, while color may be introduced by the use of tile porcelain for the inside panel on which the hammered copper hood is placed.

The view which is given below of one corner of the living room, with its fireplace, fireside seat and glimpse of the stairway behind, suggests an interesting method of handling the woodwork, and gives a general impression of the treatment of the various structural features of the interior. As is usual in a Craftsman house, these possess decorative quality without being in the least elaborate, and while they add to the beauty of the rooms they also help to minimize the task of furnishing.

The kitchen, which is a convenient size, communicates with the dining room through the pantry, which is built out between the two corner porches at the rear. A door from the kitchen also opens upon the adjacent porch.

The second floor is conveniently arranged with three bedrooms, ample closets and a large storeroom, the closets being built under the roof and not being full height except at the front. The storeroom and the large closets against the outer walls are lighted by small windows. The staircase is also well lighted.

ONE CORNER OF THE LIVING ROOM IN CEMENT COTTAGE NO. 118, SHOWING ARRANGEMENT OF THE CRAFTSMAN FIREPLACE-FURNACE, WITH BUILT-IN SEAT AND STAIRWAY BEHIND IT.

INEXPENSIVE CEMENT CONSTRUCTION FOR SUMMER AND WEEK-END ONE-STORY BUNGALOW

ALTHOUGH so simple in construction that the owner can assist in building it, this little bungalow will prove a well-planned, serviceable and attractive dwelling. The walls sunshine to each one of the five rooms within.

In spite of the extreme simplicity of its construction and lay-out of the interior—or perhaps we should rather say because of this

Published in The Craftsman, December, 1909.

INEXPENSIVE CEMENT BUNGALOW: NO. 80.

and partitions are of cement mortar upon metal lath. The girders of the house are supported upon concrete piers, less expensive than a stone foundation. The base of the chimney runs to the depth of the piers. The porch floor may be of cinder concrete, the same as used for sidewalks, slightly slanted so that it will drain easily, and the porch roof supports are of logs. The rafters are sheathed with V-jointed boards, dressed, and finished on the under side. These boards make the only ceiling to the cottage, and above them are laid strips of Ruberoid roofing. Within, all the structural beams are left exposed and are smoothed and stained.

The big chimney in the living room contains also the flue of the kitchen range. Besides these two rooms there are two bedrooms, a bathroom and many convenient closets, the arrangement, as the floor plan shows, being both compact and convenient.

The groups of windows with their small square panes not only add a touch of interest to the plain cement walls of the building, but give ample air and simplicity—the bungalow, when comfortably and tastefully furnished, should make a very charming little summer home, and would certainly permit a minimizing of all housework.

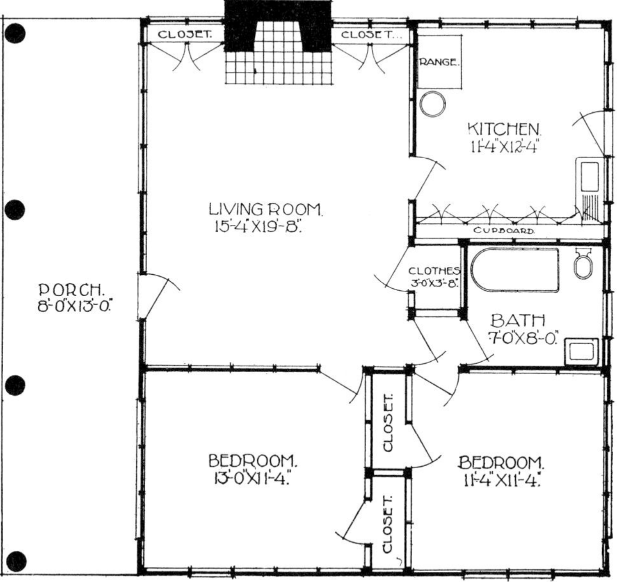

FLOOR PLAN OF CEMENT BUNGALOW: NO. 80.

A CRAFTSMAN CITY HOUSE DESIGNED TO ACCOM-MODATE TWO FAMILIES

Published in The Craftsman, October, 1907.
VIEW OF FRONT AND SIDE, INDICATING THE SIMILARITY IN ARRANGEMENT OF UPPER AND LOWER FLOORS.

SOME little time ago a problem was brought to us which proved interesting, not only in itself but on account of its application to a condition which in city life is almost universal. It was this: A man living in Brooklyn, who owned a lot thirty feet wide by a hundred feet deep, desired to build within this space a Craftsman house which should not only show a departure from the usual design of the city house in such matters as economy of space, arrangement of rooms, and interesting structural features that would serve as a basis for interior decorations and furnishing, but would accommodate two families who desired to live independently of one another, as they would in separate houses.

It had often been brought to our attention by people living in cities that most of our plans were for detached dwellings in the country or the suburbs, where the houses could have the environment of ample grounds and be given all the room necessary to carry out any idea of arrangement that might seem desirable. This method of living in the open with plenty of room and green growing things all around has always been so much more in accordance with the Craftsman idea of a home environment than any house cramped to fit the dimensions of a city lot, that our suggestions for house building have as a rule naturally taken the form of dwellings best fitted for the country. The number and frequency, however, of the requests which have come to us from time to time for city houses made the problem shown here one that we took much interest in working out.

As the owner desired a detached house with a walk on either side it was necessary to bring the dimensions of our plan within a very narrow space. Accordingly the width of the

45

A CRAFTSMAN CITY HOUSE DESIGNED FOR TWO FAMILIES

house was fixed at twenty-five feet, with a depth of sixty-eight feet, including a front porch nine feet in width. The first story is occupied by a tenant, the owner reserving the second floor for himself and his family.

It will be noticed by looking carefully at the floor plans that only the front porch, the vestibule and the rear entrance can be used in common by both families. There is no connection between the two apartments. One door from the vestibule opens to the stairway which leads to the second story and the other opens into the hall of the first story. Both stories are the same in arrangement and are planned to secure the greatest possible open-

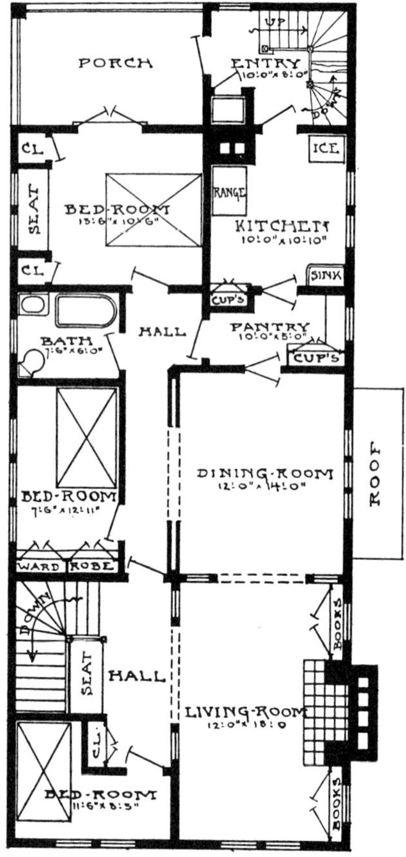

SECOND FLOOR PLAN.

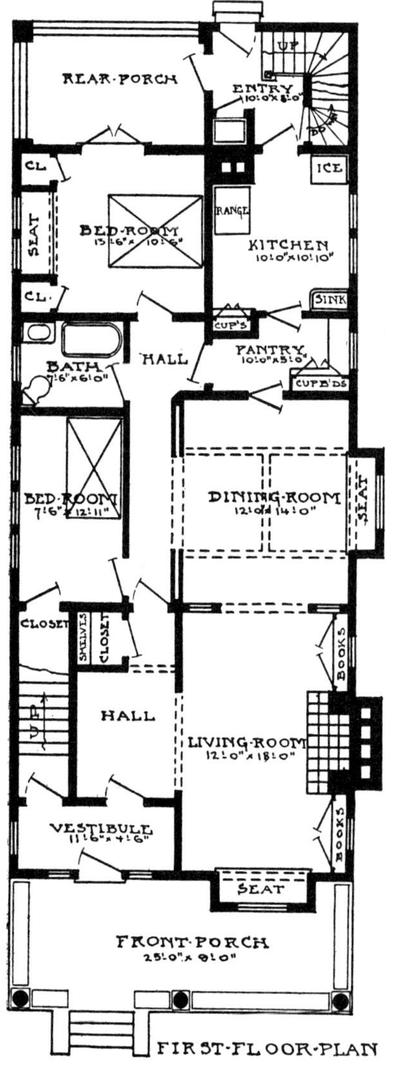

FIRST·FLOOR·PLAN

ness and freedom of space in the living rooms. The large bedrooms at the back of the house open upon rear porches, which are glassed in for the winter and screened in summer to serve as outdoor sleeping rooms.

The floor plans themselves give the best idea of the arrangement of space in the apartments. Both kitchens are provided with gas stoves and individual boilers for hot water. A dumbwaiter runs from the cellar to the attic for the convenience of the upper apartment. The cellar contains individual store rooms and coal bins, and a big laundry with a set of three tubs and a stove was installed, together with a hot water heating system for the entire house. The attic is divided in a way that provides two rooms in the dormer for the servants of both apartments, as well as a large room facing the front that can be used as a dry room in inclement weather or as a play-room for children. The cellar walls are of concrete faced with split field stone.

CITY HOUSE WITH INTERESTING FACADE AND SLEEPING BALCONY, AND HOMELIKE INTERIOR

WE are showing here a house of Craftsman construction with open, simple interior and yet adapted to a restricted space and suited to life in the city. The house is planned to be built on a long, narrow lot, 25 feet wide. As the floor plans show, the building is semi-detached, having on one side a party wall, while on the other the windows are arranged to overlook a side court or alley. The arrangement of the rooms, therefore, is considerably modified by the limited space allowed and the necessarily long, narrow shape of the building, and the exterior is modified to an even greater degree, because, as the house is not intended to be built on a corner lot, the façade is all that can be seen from the street.

As it was manifestly impossible to introduce any of the features that make up the beauty and comfort of a country house, we have sought a new expression of our basic architectural principles.

Richly-colored, rough-surfaced Tapestry bricks are used, laid in darkened mortar with wide joints. The main roof, the dormer roof, the hood over the entrance door and the upper part of the pilasters are all dull green matt-glazed tile. The porch is screened by flower-boxes.

Published in The Craftsman, October, 1910.

THREE-STORY CRAFTSMAN HOUSE FOR THE CITY, WITH SLEEPING BALCONY AND HOMELIKE INTERIOR: NO. 99.

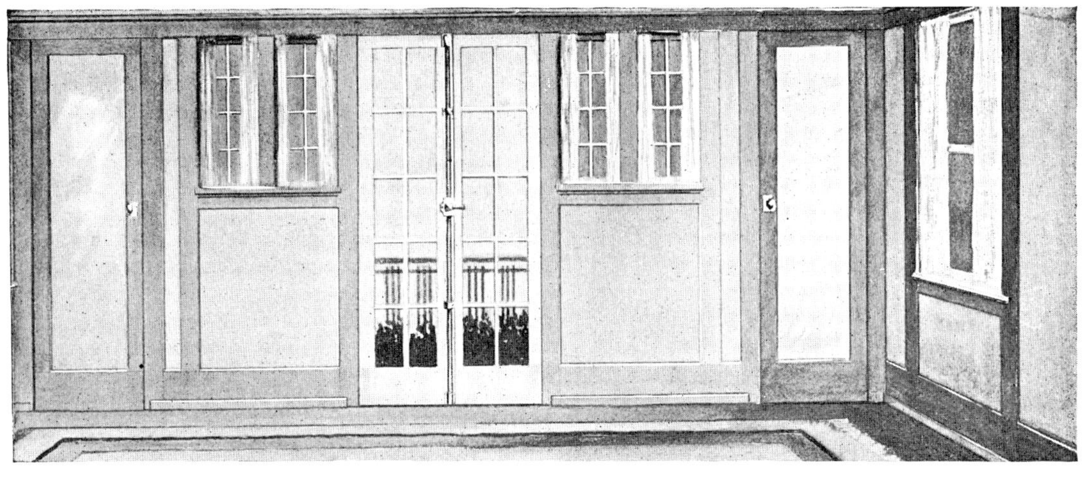

ONE END OF A BEDROOM IN HOUSE NO. 99, SHOWING TREATMENT OF WOODWORK, DOORS AND WINDOWS.

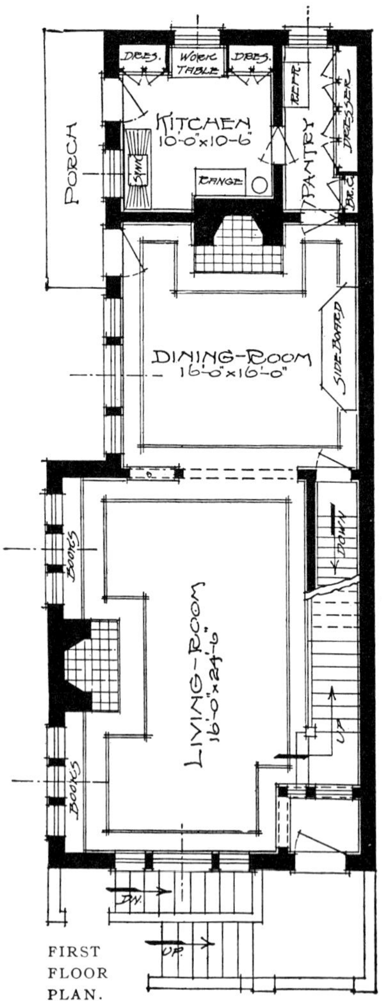

FIRST FLOOR PLAN.

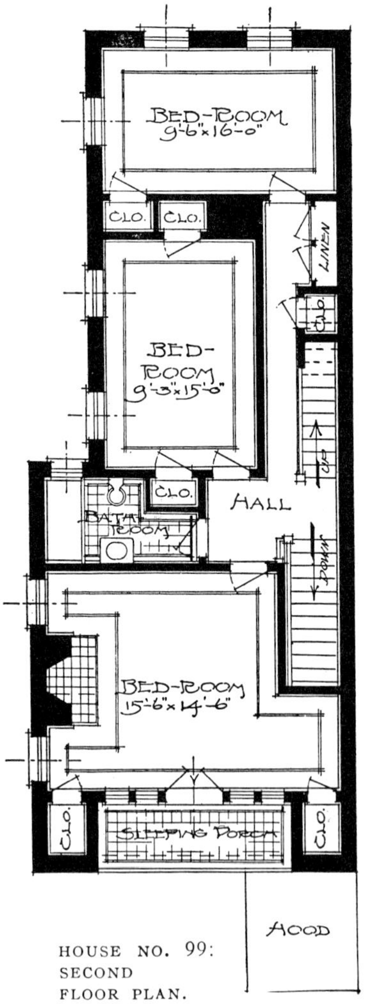

HOUSE NO. 99: SECOND FLOOR PLAN.

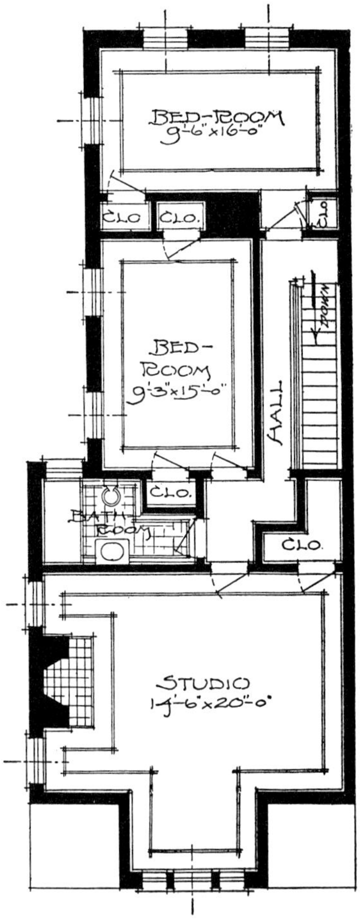

HOUSE NO. 99: THIRD FLOOR PLAN.

CRAFTSMAN CITY HOUSE WITH SECOND-STORY PORCH AND THIRD-STORY SLEEPING BALCONY

Published in The Craftsman, October, 1910.

CRAFTSMAN BRICK HOUSE FOR THE CITY, WITH ELEVEN ROOMS AND TWO SLEEPING PORCHES: NO. 100.

ELEVEN-ROOM CITY HOUSE, WITH TWO SLEEPING PORCHES

LIKE the house shown on pages 91 and 92, this building is also planned for a 25-foot city lot. But the restrictions in this case are even greater, for instead of facing an alley on one side the central portion of this house obtains light and ventilation from an interior court or air shaft on which the windows of library, dining room, bathroom and several bedrooms open. Notwithstanding this fact, the careful planning of the interior insures light and air for all the rooms.

Plain red hard-burned brick is used for the facade, the uniform dark red being varied by the darker purplish tones of the arch brick which are introduced wherever they will be effective. The roof and the hood over the entrance door are of dull red tile.

The object being to get a good design and construction as inexpensively as possible, the sleeping porch is placed on the third story just under the tiled roof. The rafters of the roof are emphasized, projecting sufficiently to give a suggestion of a pergola and affording a support for vines that might be grown in the flower-boxes.

The half-tone illustration of the staircase in the living room shows the way in which the stairs are screened by a high-backed seat and an arrangement of slats above. The fireplace in the large back bedroom is also shown, with

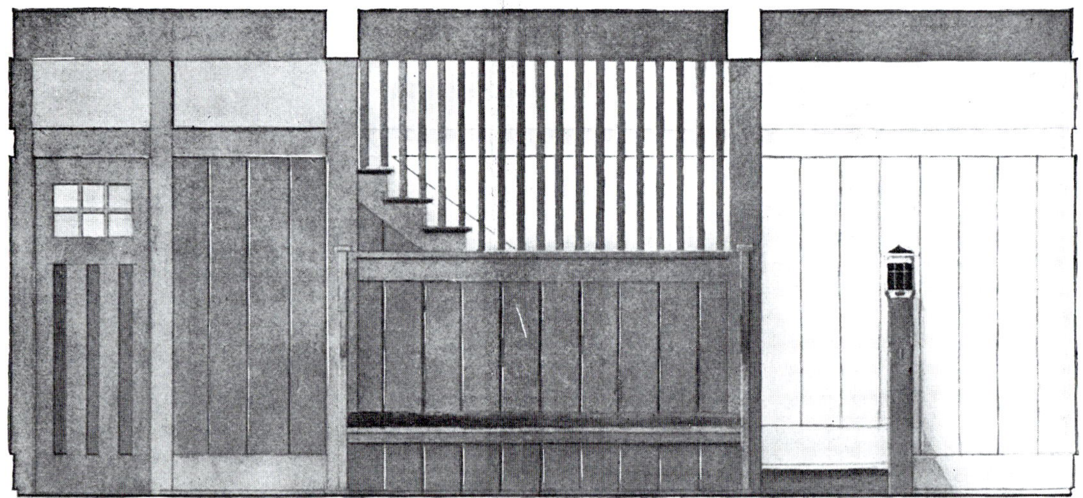

TWO VIEWS OF INTERIOR OF CITY HOUSE, NO. 100. THE FIRST SHOWS STAIRWAY GOING UP FROM LIVING ROOM, WITH BUILT-IN SEAT AND SPINDLES ABOVE; THE SECOND SHOWS FIREPLACE END OF ONE OF THE BEDROOMS, WITH GLASS DOORS ON EACH SIDE LEADING TO SLEEPING PORCH.

ELEVEN-ROOM CITY HOUSE, WITH TWO SLEEPING PORCHES

the glass doors on either side opening upon the rear porch.

The floor plans show the front entrance door opening directly into the living room, which has a large open fireplace on one side. By placing corner seats or long settles beside the hearth a very comfortable inglenook could be formed. Beyond this room is the library, with a window group in one wall overlooking the interior court, and on the opposite side are built-in bookshelves with windows set high in the wall above. Beyond the library is the dining room with its open fireplace, on the right

of which is a swing door leading to the conveniently arranged kitchen in the rear.

The second floor contains three bedrooms and bath, with ample closet room. On each side of the fireplace in the back bedroom are doors leading to the porch. On the third floor are four bedrooms and a bath. Here also there are plenty of closets. The two front rooms open upon the recessed sleeping porch, with its pleasant screen of flower-boxes. This porch has a partition across the center as shown, and is so sheltered from the weather that it could be used practically all the year round.

While city restrictions have not permitted a typical Craftsman dwelling, still the house reveals many possibilities for the making of a comfortable and homelike interior.

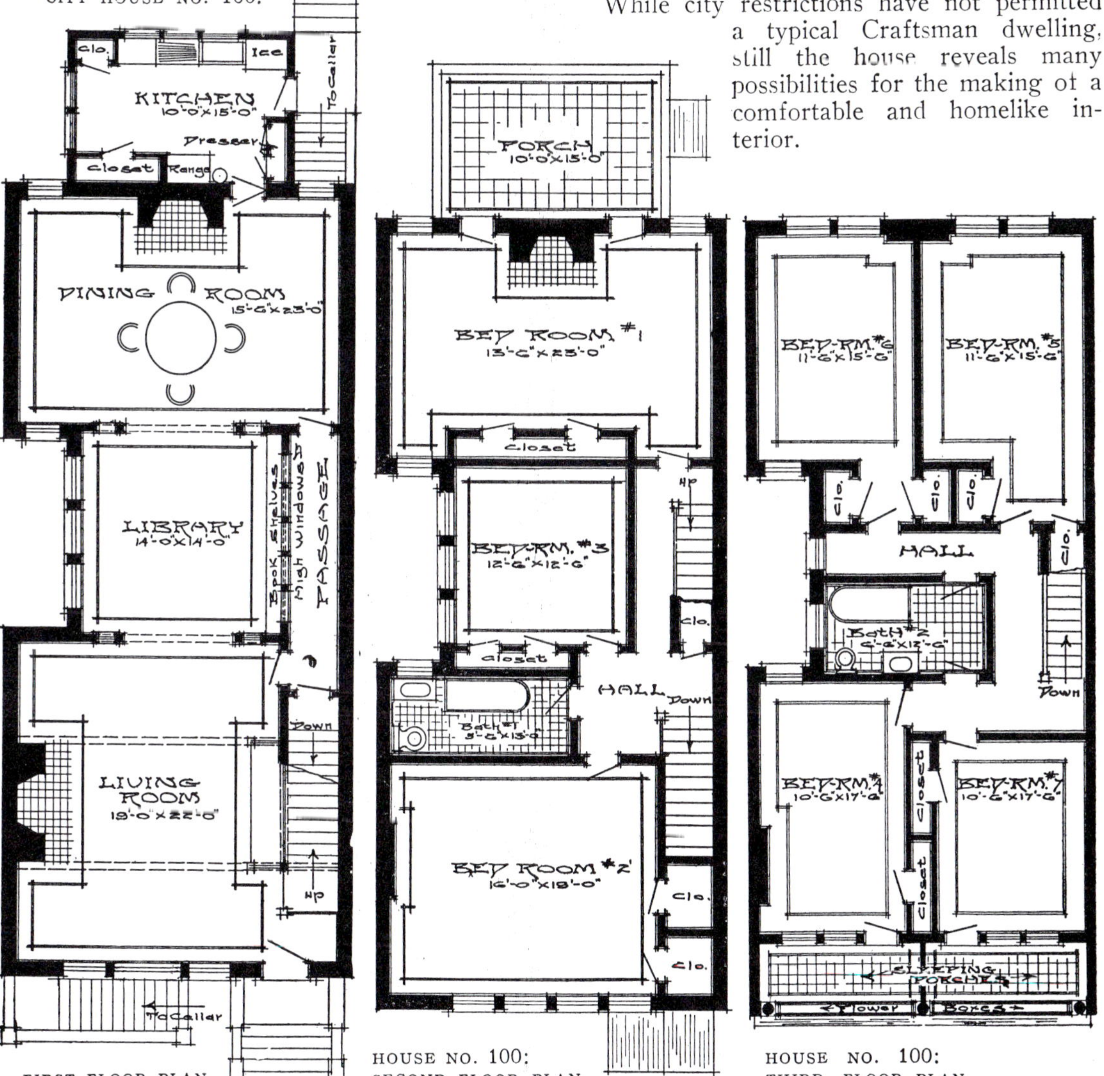

A CRAFTSMAN FARM HOUSE THAT IS COMFORT-
ABLE, HOMELIKE AND BEAUTIFUL

Published in The Craftsman, June, 1906.

VIEW FROM THE FRONT, SHOWING DORMER, GABLE AND RECESSED PORCH.   NOTE THE EFFECT OF THE BROAD LOW PROPORTIONS, THE GROUPING OF WINDOWS AND THE DECORATIVE USE OF TIMBERS.

A HOMELIKE AND BEAUTIFUL CRAFTSMAN FARMHOUSE

IF there is any one style of house that we enjoy planning more than others, it is a farmhouse,—a home that shall meet every practical requirement of life and work on the farm, and yet be beautiful, comfortable and homelike. This is our first farmhouse and we endeavored to make it characteristic in design, plan, decoration and the materials used for building. As a rule, we do not advocate the use of clapboards for sheathing the walls of a frame house, for the reason that the small, thin, smoothly planed and painted boards generally used for this purpose give a flimsy, unsubstantial effect to the structure and a characterless surface to the walls. However, clapboards are often preferred, especially in building a farmhouse, and it is quite possible to use them so that these objections may be removed. In this building the clapboards are unusually broad and thick, giving to the walls a sturdy appear-

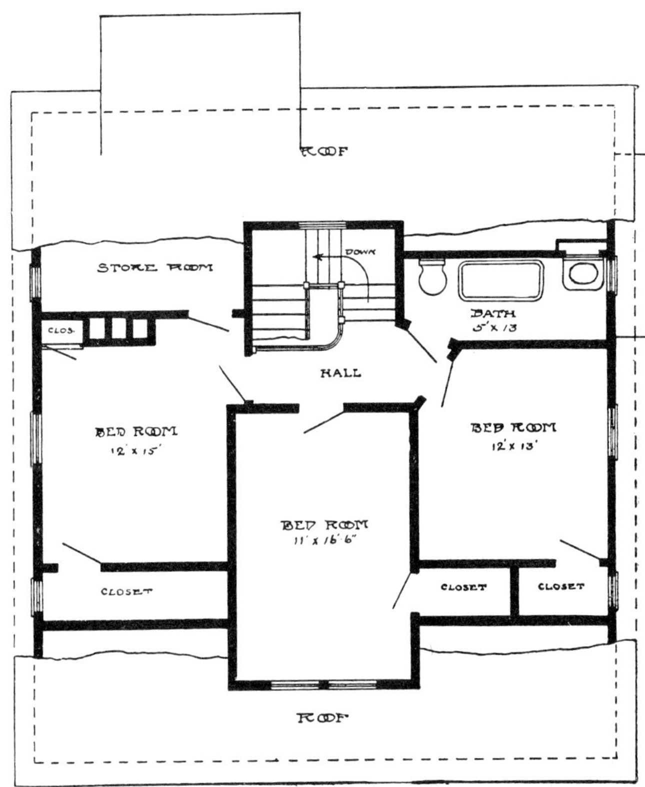

SECOND FLOOR PLAN.

FIRST FLOOR PLAN.

ance of permanence. They may be of pine, cedar, or cypress, and may be stained or painted according to individual taste and the character of the environment. If the house is to be rather dark and quiet in color, the boards might be given a thin stain of moss green or brown; or a delightful color effect may be obtained by going over the boards with a wash of much diluted sulphuric acid. With either one of these colors a good effect would be obtained by painting the timbers of the framework a light cream so that the structural features are strongly accented.

We regard this house as having in a marked degree the comfortable and inviting appearance which seems so essentially to belong to a home,—particularly to a farm home. It is wide and low, with rather a shallow pitch to the broad roof, the line of which is unbroken by the large dormers set at different

CORNER OF LIVING ROOM, SHOWING TREATMENT OF WALL SPACES BY A VERY SIMPLE USE OF THE WOODWORK, WHICH IS USED TO GIVE THE EFFECT OF A BROAD, PLAIN FRIEZE. NOTE THE MANNER IN WHICH THE WINDOW AND DOOR FRAMINGS ARE RELATED TO THE LOWER BEAM OF THIS FRIEZE.

heights. The entrance porch, which is of ample size, is recessed to its full width. The timbers which accent the construction give special interest to the interior, as they are so placed as to add to the apparent width of the house, and are arranged so as to avoid, by means of the prominent horizontal lines of the beams, any possible "spotty" effect which might result if the vertical lines of the framework were not so relieved. This device is especially apparent in the grouping of the three windows which light the gable. The plan of the house makes it necessary that these be rather far apart, but they are built together by the beams so as to form a symmetrical group rather than to give the impression of three separate windows in a broad wall space. The same effect is preserved throughout the lower story by the massive beam which extends the entire width of the house, not only defining the height of the lower story but serving as a strong connecting line for the window and door framings which all spring from the foundation to the height of this beam.

A small vestibule, which serves to cut off draughts that might come from the entrance door, opens into the central hall which forms a connecting link between the living room on one side and the library and dining room on the other. The staircase, which is opposite the entrance, is placed well toward the back

A HOMELIKE AND BEAUTIFUL CRAFTSMAN FARMHOUSE

CORNER OF DINING ROOM SHOWING SIDEBOARD BUILT INTO A RECESS, WITH GROUP OF WINDOWS AND DISH CUPBOARD.

of the house, giving as much width as possible to the hall. A small coat closet occupies a few feet of space that has been made available between the vestibule and the living room, so that the lines of both hall and living room are uninterrupted.

The living room has the advantage of every ray of sunshine which strikes that side of the house, as it is not sheltered by the porch. It is quite a long room in proportion to its width and the entire end at the rear is taken up by the fireplace and the two seats which, extending from it at right angles, give the effect of a deeply recessed fireside nook. A single chimney is made to do service for the entire house, as it is arranged to accommodate three flues.

A ROOMY, HOMELIKE FARMHOUSE FOR LOVERS OF PLAIN AND WHOLESOME COUNTRY LIFE

FRONT VIEW SHOWING PORCH, DORMER AND SLEEPING BALCONY.

BOTH in exterior seeming and in interior arrangement and finish, this building is essentially a farmhouse,—not of the comfortless type that we have been accustomed to of late years, but one that is reminiscent of earlier days, when a farmhouse was in very truth the homestead and as such was large, substantial, comfortable and inviting. The design is very simple, with clapboarded or shingled walls and a broad sheltering roof, the straight sweep of which is broken by a large dormer on either side. The wide veranda in front is recessed, forming a sheltered porch that could be used for much outdoor life. The windows as suggested here are all casements, those on the upper story being protected from the weather by the broadly overhanging roof and the lower ones sheltered by hoods. At the front of the house the dormer is extended to form a good-sized sleeping porch and at the back it accommodates the bathroom.

As the general effect of the house is broad and low, it is fitting that very little of the foundation should be visible. A far better effect is given if no attempt is made to establish too strict a grade line, as the house seems to fit the

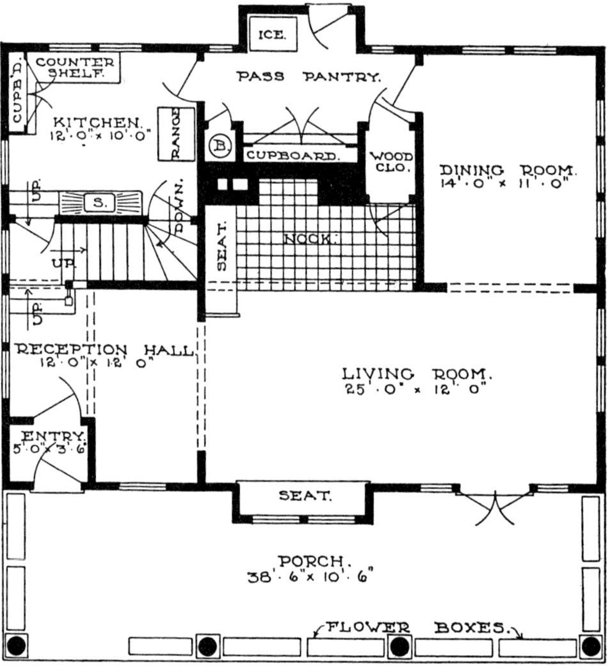

FIRST STORY PLAN

ground much better if the foundation is accommodated to the natural irregularities and if the floor of the porch is very little elevated above the turf.

The interior arrangement, while simplicity itself, is very convenient. There is hardly anything to mark the divisions between the reception hall, living room and dining room, so that these names rather serve to indicate the uses to which the different parts of this one large room may be put than to imply that they are separate rooms. In the very center of the house is the large fireplace nook which naturally forms the center of interest and attraction, with its ample chimneypiece of the split field stone and the comfortable fireside seat beside the hearth. Were it not for the arrangement of this large open space, there might be a sense of bareness; but this is entirely obviated by the shape of the room, the prominence given to the fireside nook, and the liberal use of wood in the form of beams, wainscots, seats and such built-in fixtures as may be necessary.

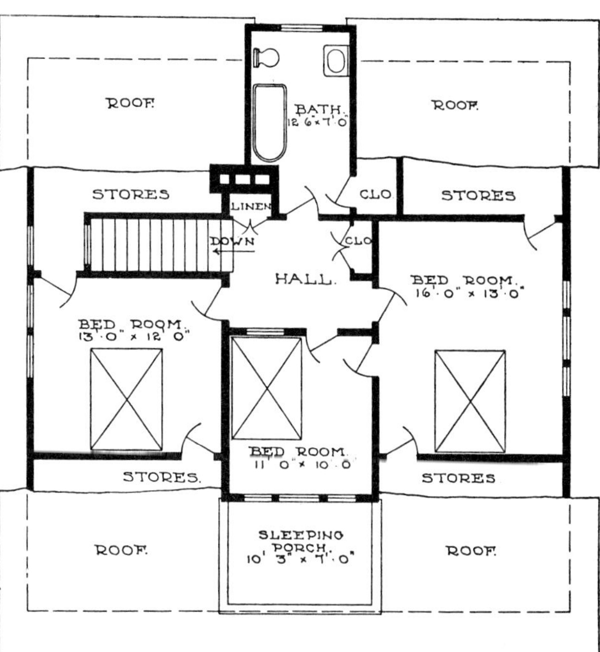

SECOND STORY PLAN.

FIRESIDE NOOK, GIVING AN IDEA OF THE BROAD CHIMNEYPIECE BUILT OF SPLIT FIELD STONE AND OF THE **FIRESIDE** SEAT, WHICH IS MADE OF WIDE BOARDS V-JOINTED.

A FARMHOUSE DESIGNED WITH A LONG, UN-BROKEN ROOF LINE AT THE BACK

Published in The Craftsman, January, 1909.

FRONT VIEW, SHOWING RUSTIC PERGOLA AND INTERESTING CONSTRUCTION THAT SUPPORTS THE OVERHANG.

REAR VIEW SHOWING WIDE SWEEP OF ROOF AT THE BACK IN PLACE OF THE CUSTOMARY "LEAN-TO."

A FARMHOUSE WITH A LONG ROOF LINE

WE feel that the design for this farmhouse is one of the most satisfactory that we have ever done, not only because the building, simple as it is, is graceful in line and proportion, but because the interior is so arranged as to simplify the work of the household and to give a good deal of room within a comparatively small area.

The plan is definitely that of a farmhouse, and in this frank expression of its character and use lies the chief charm of the dwelling. The walls might be covered with either shingles or clapboards, according to the taste and means of the owner. If the beauty of the building were more to be considered than the expense of construction, we should recommend the use of rived cypress shingles, as these are not only very durable but have a most interesting surface. The only difficulty is that they cost about double the price of the ordinary shingles. As the construction of the house in front is such that a veranda

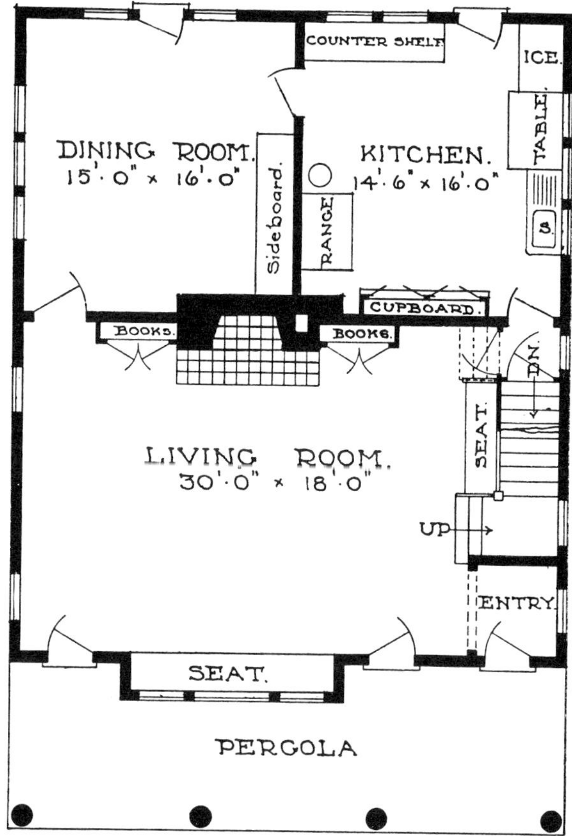

FIRST STORY FLOOR PLAN.

SECOND STORY FLOOR PLAN.

would be rather a disfigurement than an improvement, we have supplied its place by a terrace covered with a pergola. The terrace would naturally be of cement or vitrified brick and the construction of the pergola should be rustic in character. One great advantage of such a pergola is that the vines that cover it afford sufficient shade in summer, while in winter there is nothing to interfere with the air and sunlight, which should be admitted as freely as possible to the house. We have allowed the roof to come down in an unbroken sweep toward the back because of the beauty and unusualness of this long roof line as compared with the usual square form of a house with the lower roof of a porch or lean-to at the back. Furthermore, by this device there is considerable space for storage left over the kitchen and dining room. The entry opens into the living room at right angles with the entrance door and this opening might be curtained to avoid draughts.

59

SMALL CRAFTSMAN FARMHOUSE OF STONE AND SHINGLES, SIMPLE AND HOMELIKE IN DESIGN

VILLAGE or open country would be the environment most suitable for this little house. The walls are shingled and the low foundation is of field stone, sunk into a site that has not been too carefully leveled off. This irregularity of the ground is utilized in a practical way, the slope at the back being sufficient to allow space for the cellar windows, while at the front it is high enough to bring the cement floor of the porch almost upon a level with the lawn. Instead of parapets, the spaces between the pillars of the porch are filled with long flower-boxes which serve as a slight screen and add a note of color to the house. The roof extends over the porch and the sweep of it is broken by the dormer with its group of casements which give light to both bedrooms and the sewing room.

The entrance door at the corner of the porch opens directly into a little nook in the living room. Directly opposite the door is the stairway which runs up three steps to a square landing and then turns and goes up behind the wainscoted wall of the room. The whole wall on this side is taken up by the long fireside seat. The chimneypiece of split field stone occupies the space between the wall and the opening into the dining room. This little ingle-nook is shown in the illustration which gives some idea of the treatment of the wainscot, posts and beams. A decorative note is added by the Craftsman hanging lanterns above. Behind the dining room is a small, conveniently arranged kitchen.

Upstairs there are two bedrooms, a tiny sewing room, bathroom and hall. Both bedrooms communicate with the sewing room which is placed between them and is provided with a wardrobe. There is ample closet room at each corner beneath the slope of the roof, and one of the bedrooms also has a closet with shelves against the inner wall beside the central chimney. Seats are built into the front recesses beneath the windows.

The house is a small one, having only five rooms and bath, but the compactness of its arrangement and the sense of space given by the openness of the lower floor plan result in a very homelike interior.

For the farmer who is planning to build a home of his own, the farmhouse shown here, as well as those on other pages of the book, should prove full of practical and helpful suggestions in arrangement and design. We have endeavored to keep the plans as simple and inexpensive as possible, and at the same

STONE FIREPLACE IN CORNER OF LIVING ROOM OF FARMHOUSE NO. 61, WITH FIRESIDE SEAT AND GLIMPSE OF STAIRCASE.

time to secure the greatest possible comfort and beauty, both in the exterior of the building and the rooms within. For the farmer has tolerated too long the discomfort and bareness of the average farmhouse. He needs a home for himself and his family which is both comfortable and cheerful, a place where he may find rest and recreation after the day's work, and in which the necessary labor of the household may be done under conditions which make it as light as possible.

When our farmhouses are designed from this standpoint of utility and beauty, we shall no longer regard the work within their walls as a round of drudgery, a necessary evil. Instead, the so-called "menial" tasks of cooking, sweeping, sewing, will be a source of pleasure and pride, and in place of the old weary attitude toward work we shall find ourselves laboring with interest, with enthusiasm,—qualities which are inevitable in the building up of the ideal home.

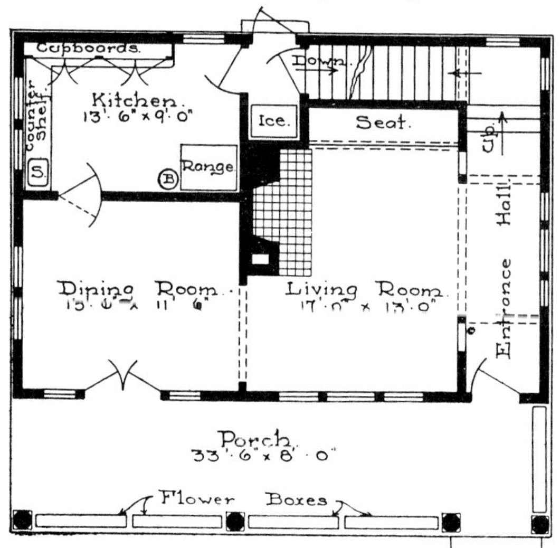

FARMHOUSE NO. 61: FIRST FLOOR PLAN.

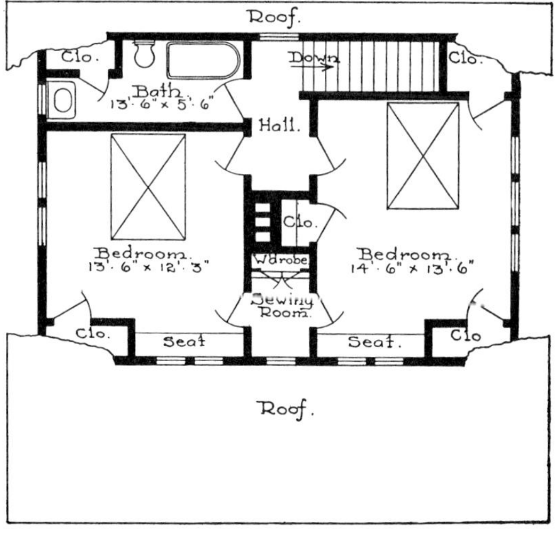

FARMHOUSE NO. 61: SECOND FLOOR PLAN.

COMFORTABLE, CONVENIENT, HOMELIKE FARM-HOUSE WITH CONNECTED WOODSHED AND BARN

Published in The Craftsman, February, 1911.   FARMHOUSE WITH WOODSHED AND BARN: NO. 108.

BUILT on a stone foundation, with walls of sawn shingles, with roof of Ruberoid, and hewn log posts supporting the roofs of pergola, porches and balcony, this farmhouse presents a simple but attractive exterior. The building is planned especially for convenience and economy of labor, and is heated and ventilated by a Craftsman fireplace-furnace. Coal and wood closets are provided where fuel can be stored, easily accessible to fireplace and kitchen. A summer kitchen is also provided containing stove and laundry tubs, while an outdoor dining room, its long table and benches enclosed from the yard by a curved hedge, forms a most charming place for serving meals during the summer.

ONE SIDE OF LIVING ROOM IN FARMHOUSE NO. 108, SHOWING RECESSED FIREPLACE AND EFFECTIVE SIMPLICITY OF WALL TREATMENT.

FARMHOUSE WITH CONNECTED WOODSHED AND BARN

The woodshed provides a passage under shelter to the barn and sufficiently isolates the barn from the house to remove any objectionable features. The barn is not intended to accommodate much stock, but a box stall and one single stall have been planned for horses, and a separate room large enough for three or four cows has been partitioned off with a solid wall. This stall has an outside entrance.

We located the feed bins in the loft and convey the feed to the first floor through metal chutes. A hay chute is also provided. Ample room for carriage, wagon and farm tools is arranged for on the first floor. The corn crib is constructed of slats as shown; this should be lined on all sides, top and bottom, with a fine mesh wire to keep out rats or mice.

One of the great advantages of the fireplace-furnace being located on the first floor is the fact that there is no heat in the cellar. Fruit and vegetables can be stored in the cellar and will keep nicely all the winter.

The second floor plan shows a very simple arrangement. There are four bedrooms of convenient size, and a bathroom, all opening out of the small central hall. Plenty of closets are provided. The two bedrooms at the right have glass doors lead-ing to the sleeping porch which runs across the side of the house and is sheltered by the roof extension.

The illustration of the living room suggests a simple but interesting treatment of woodwork and walls, and the use of small panes in the windows always adds a decorative note to the interior. The recessed hearth is a somewhat unusual feature, and increases the homelike air of the long cheerful room.

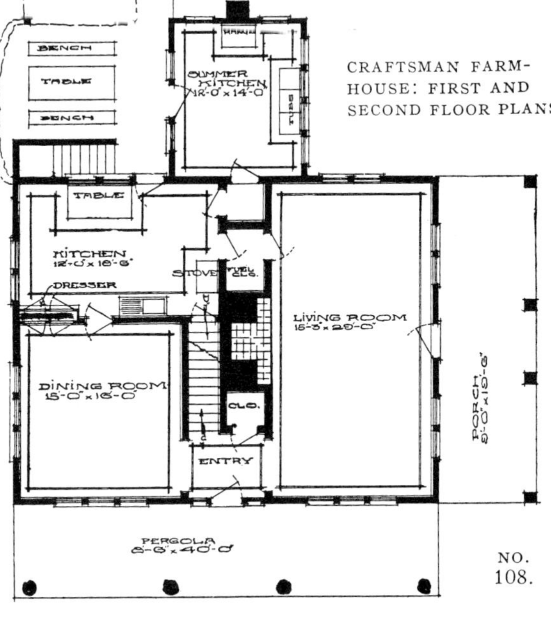

CRAFTSMAN FARMHOUSE: FIRST AND SECOND FLOOR PLANS.

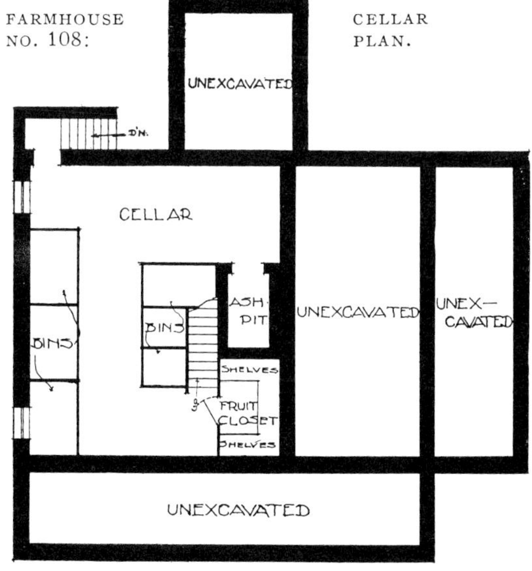

FARMHOUSE NO. 108: CELLAR PLAN.

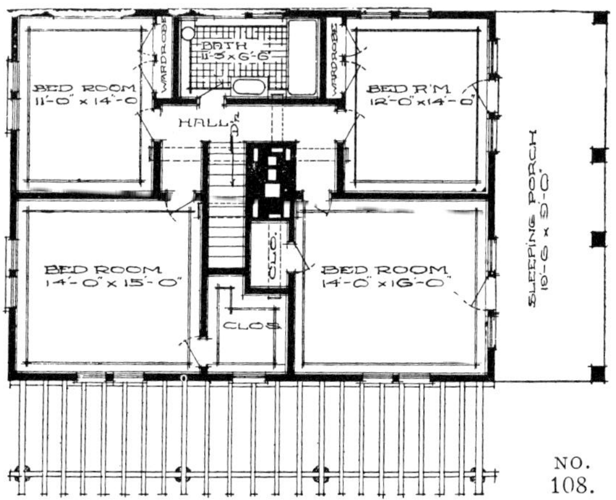

NO. 108.

HOUSE WITH COURT, PERGOLAS, OUTDOOR LIVING ROOMS AND SLEEPING BALCONIES

HOUSE DESIGNED FOR OUTDOOR LIFE IN A WARM CLIMATE.

LIFE in a warm country, where there is much sunshine and where it is possible to be out of doors during the greater part of the time, was specially taken into consideration in the designing of this house, for the plan makes as much account of the terraces, porches and the open paved court as it does of the rooms within the walls of the building. Such a plan would serve admirably for a dwelling in California or in the Southern States, but would be advisable only for specially favored spots in the North and East, as its comfort and charm necessarily depend very largely upon the possibility of outdoor life.

As originally planned, the walls of the lower story are to be built of cement or of stucco on metal lath. The upper walls are shingled. The roof is of red tile and the foundation and parapets are of field stone. As with all these houses, though, the materials used are entirely optional and can be varied according to the taste of the owner, the requirements of the landscape or the limitations of the amount to be expended, as the building would look quite as well if constructed of concrete or of brick, and with clapboards in the place of shingles. If a wooden house should be preferred, the walls from top to bottom could either be shingled or sheathed with wide clapboards, while the roof is equally well adapted to tiles, slates or shingles. The first of the perspective drawings gives a view of the whole house as seen from the rear, showing the pergola at the back and the design of the roof, which we consider specially attractive. The second drawing shows the side of the house instead of the front, as by taking this view it is possible to include both porch and court and also show the balcony and outdoor sleeping room on the upper story. A broad terrace runs across the front of the house and continues around the side, where it forms a porch which is meant to be used as an outdoor living room. This porch is nearly square in shape and is either tiled with Welsh quarries or, if a less expensive flooring be desired, is paved with red cement marked off into squares that measure about nine inches each way. This floor has a close resemblance to one made of Welsh quarries and is dry and durable. In flooring a porch of this kind it is always better to avoid the use of plain brick, as this porous material gathers and holds moisture to such an extent that the floor is seldom dry.

HOUSE WITH COURT, PERGOLAS AND OUTDOOR ROOMS

The entrance door opens from this porch into the hall, which is separated from the living room only by two panels open at the top after the usual Craftsman fashion, the wood running only a little above the height of the two bookcases, which may either be built in or movable, as desired. Directly opposite this entrance is the large fireplace, which is recessed so as to form a fireside nook. Seats are placed on either side and the tiled hearth extends the full length of these. Back of them, in the small recesses left on either side of the fireplace, are built-in bookcases with casement windows set above. A square bay window, below which is a broad window seat, looks out upon the terrace, and double glass doors from both living room and hall bring this part of the house into very close communication with the outside world; an important feature in the planning of a house intended for life in a warm climate where there is little rain.

The dining room has every appearance of being merely a large square recess in the living room, as the division between them is only indicated and the dining room is just large enough to afford comfortable accommodation for a good-sized dining table and the necessary furniture. The sideboard, which is built in, occupies the entire end of the room and a group of three casement windows are set in the wall just above it.

The floor plan shows the convenient arrangement of the hall, staircase and closets, everything being grouped within a small compass so that not an inch of space is wasted. The arrangement of pantry and kitchen is equally convenient and plenty of cupboard room is provided for dishes and the necessary kitchen utensils.

The chimney that is used for the kitchen range has space also for a flue leading from the fireplace on the porch outside. We are greatly in favor of these outdoor fireplaces, because there are many days and evenings when it is almost warm enough to stay out of doors, and yet without a fire it is not quite comfortable. Also, a fire in the open air has always something of the charm of a camp fire. The placing of this one is peculiarly

desirable, as it not only makes a pleasant sitting room of the porch, but also has much of the charm of a garden, as from the porch one steps down into the court, which is surrounded on the outside by a vine-covered pergola and which may be paved or not, as desired. Even when these courts are paved they often hold growing trees or a fountain, so that both shade and the nearness of green, growing things are possible, while the court itself seems merely an extension of the porch. The den, which can be closed off by doors from the rest of the house in case privacy is desired for work or reading, has double doors leading to the square entrance porch and also to the court.

On the second floor there are three large bedrooms, plenty of closet room and three baths. One of these is for the exclusive use of the maids and opens from the maids' room at the back. The other two are placed so

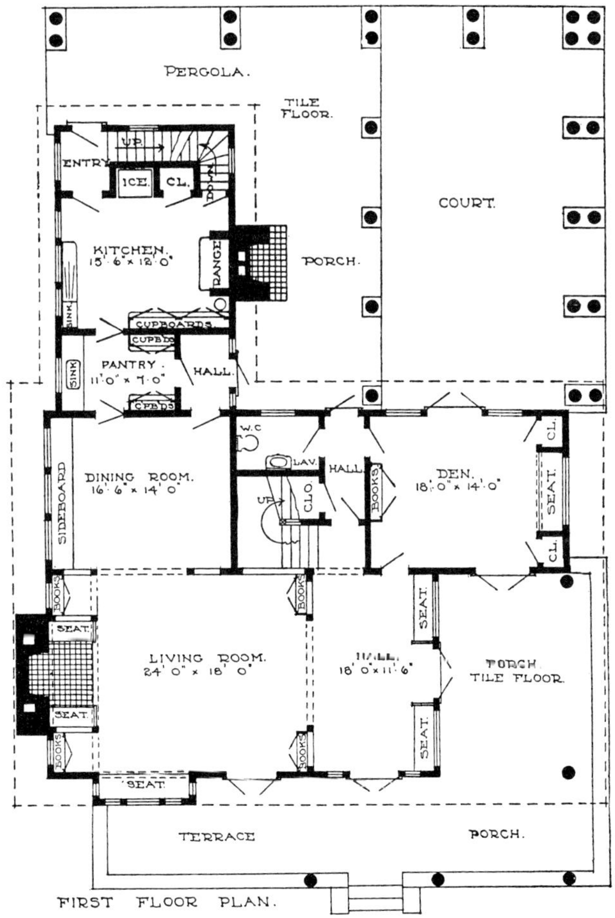

FIRST FLOOR PLAN.

CORNER OF HOUSE SHOWING PERGOLA AND SLEEPING PORCH.

that each one is accessible from two bedrooms, counting the outdoor sleeping room as one. The linen and clothes closets are so placed that they occupy the least possible amount of space. The central hall is more in the nature of a corridor running around the four sides of the staircase well, and at the back is a long window seat built beneath a group of windows that look out over the court and pergola.

The matter of interior woodwork and general scheme of color and decoration would depend very largely upon the part of the country in which the house is built. Its design is primarily that of a California house and reflects the spirit which rules the new architecture that is springing up in that country. Therefore it would seem quite in keeping to suggest that the inside of the house be finished after the well-known California style, because no other would be so completely in harmony with the plan of the exterior.

Living so much out of doors, the Californians almost instinctively make the transition between outdoors and indoors as little marked as possible by finishing the interior of their houses in the most natural way.

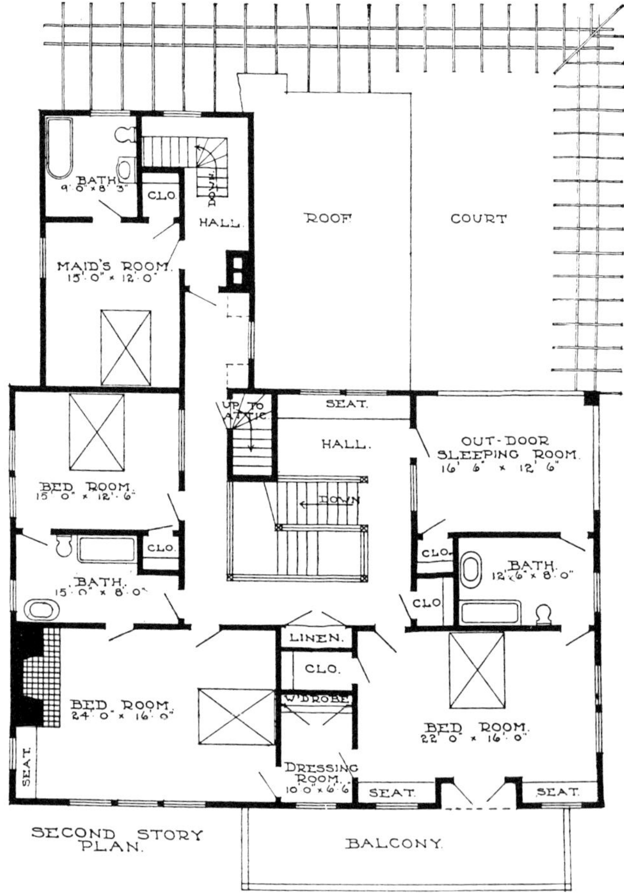

THE CRAFTSMAN'S HOUSE: A PRACTICAL APPLICATION OF OUR THEORIES OF HOME BUILDING

WHILE all the houses illustrated in this book are of Craftsman design, the dwelling shown here is perhaps the most complete example in existence of the Craftsman idea, for the reason that it is to be built by the founder and editor of THE CRAFTSMAN at "Craftsman Farms," his estate in New Jersey, and will be used there as his own home. Therefore in this case the tastes of the designer are one with the tastes and needs of the owner, who has found no creative work more absorbingly delightful than this planning of a home which he intends to live in for the rest of his life. In addition to this it affords the opportunity for working out personally, in every practical detail, all the theories which have been applied to the houses of other people.

Craftsman Farms was apparently planned by nature for the site of just such a house. It has heavily wooded hills, little wandering brooks, low-lying meadows and plenty of garden and orchard land; and the house will be built on a natural terrace or plateau half way up the highest hill. The building faces toward the south, overlooking the partially cleared hillside, which runs down to the orchard and meadows at the foot and which needs very little cultivation to develop it into a beautiful sloping greensward with here and there a clump of trees or a mass of shrubbery. There is a friendliness about the natural conformation of the land which makes it seem homelike before one stone is laid upon another or one bit of underbrush is cleared away, for the combination of sheltering hills and woods with a sheltered swale or meadowland gives interesting

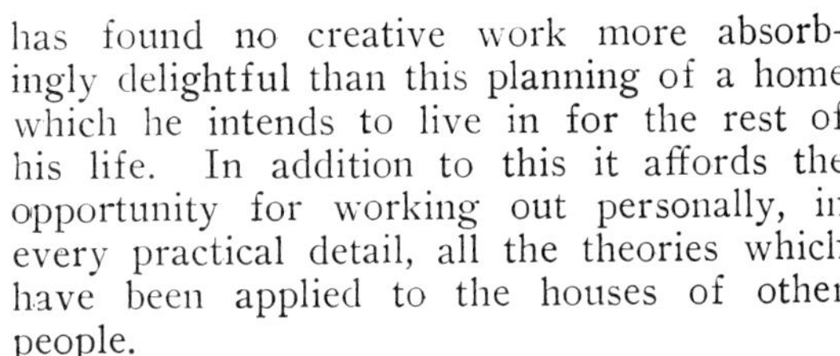

Published in The Craftsman, October, 1908.

FRONT VIEW OF THE HOUSE AT "CRAFTSMAN FARMS", SHOWING PERGOLA AND RECESSED ENTRANCE PORCH AND AT THE SIDE THE OUTDOOR DINING ROOM WITH FIREPLACE AND CHIMNEY BUILT AT THE END. THE PLASTER PANELS BETWEEN THE SEVERAL GROUPS OF WINDOWS ARE TO BE FILLED WITH LARGE PICTURE TILES SYMBOLIZING FARM LIFE AND INDUSTRIES.

variety in the immediate surroundings, while the view of the whole country from the hilltop through the gaps in the surrounding hills does away with any sense of being shut in.

In designing the house, the first essential naturally was that it should be suited exactly to the requirements of the life to be lived in it; the second, that it should harmonize with its environment; and the third, that it should be built, as far as possible, from the materials to be had right there on the ground and left as nearly as possible in the natural state. Therefore the foundation and lower walls of the building are of split field stone and boulders taken from the tumbledown stone fences and loose-lying rocks on the hillsides. The timbers are cut from chestnut trees growing on the land, and the lines, proportions and color of the building are designed with a special view to the contour of the ground upon

DETAIL OF LIVING ROOM SHOWING PIANO, PICTURE WINDOW AND BOOKCASES.

which it stands and the background of trees which rises behind it.

The hillside site, affording, as it does, well nigh perfect drainage, makes it possible to put into effect a favorite Craftsman theory,—that a house should be built without a cellar and should, as nearly as possible, rest directly on the ground with no visible foundation to separate it from the soil and turf in which it should almost appear to have taken root. The house is protected against dampness by making the excavation for the foundation down to clear hard soil, filling it in partly with the smaller pieces of stone that were rejected from the walls and placing on this a thick layer of broken stone leveled off with an equally thick layer of Portland cement and concrete, making it level and smooth like a pavement. All of this foundation is drain-tiled both inside and out. On the top of the cement floor is a double layer of damp-proofing, which extends without a break up the wall, and a thick layer of tar and sand, in which the floor timbers are bedded. Another layer of waterproof paper covers this; and then comes the floor itself—as completely protected from moisture as if it were on the top story of the building. The heating plant and laundry are provided for in a separate building and the stone storage

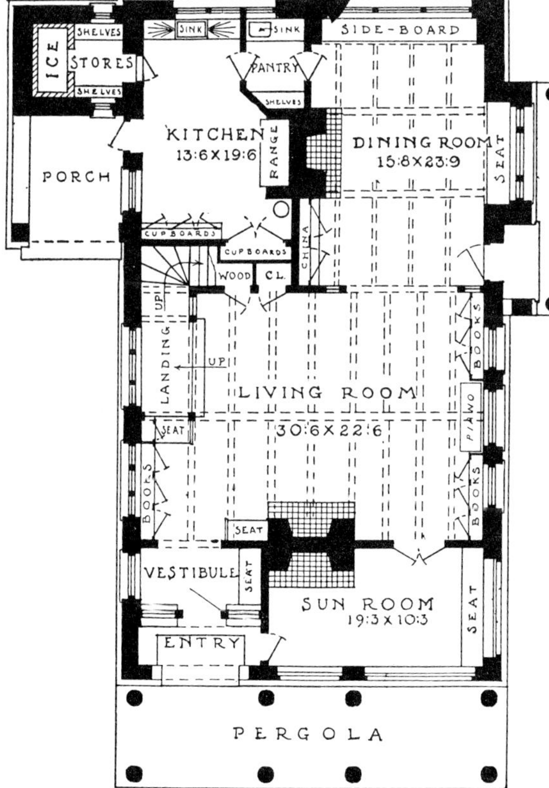

FIRST STORY FLOOR PLAN.

END OF LIVING ROOM, ILLUSTRATING HOW THE STAIRCASE WITH ITS LANDING MAY BE MADE THE PROMINENT STRUCTURAL FEATURE OF A ROOM.

vaults for vegetables and the like are sunk into the side of the hill.

No effort has been made to give the appearance of a grade line, the ground being allowed to preserve its natural contour around the stone walls of the first story. The upper walls are of plaster and half-timber construction. The plaster is given a rough pebble-dash finish and a tone of dull brownish green brushed off afterward so that the color effect varies with the irregularity of the surface. In each one of the large panels ultimately picture tiles will be set, symbolizing the different farm and village industries,—for example, one will show the blacksmith at his forge; another a woman spinning flax; others will depict the sower, the plowman and such typical figures of farm life. These tiles will be very dull and rough in finish and colored with dark reds, greens, blues, dull yellows and other colors which harmonize with the tints of wood and stone.

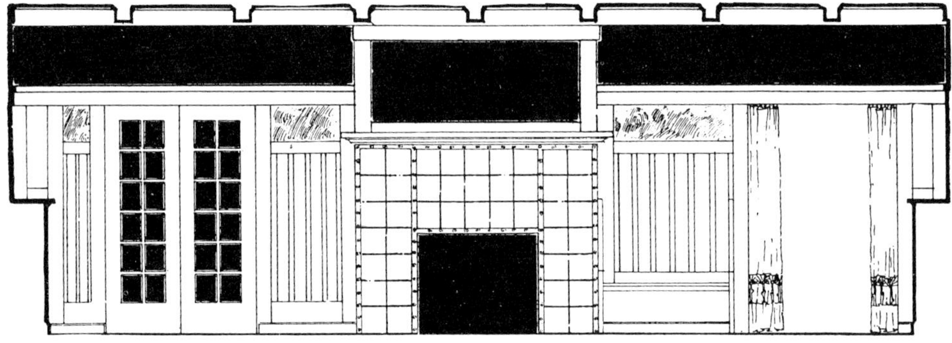

DETAIL OF LIVING ROOM SHOWING FIREPLACE, DOORS INTO SUN ROOM AND ENTRANCE TO VESTIBULE.

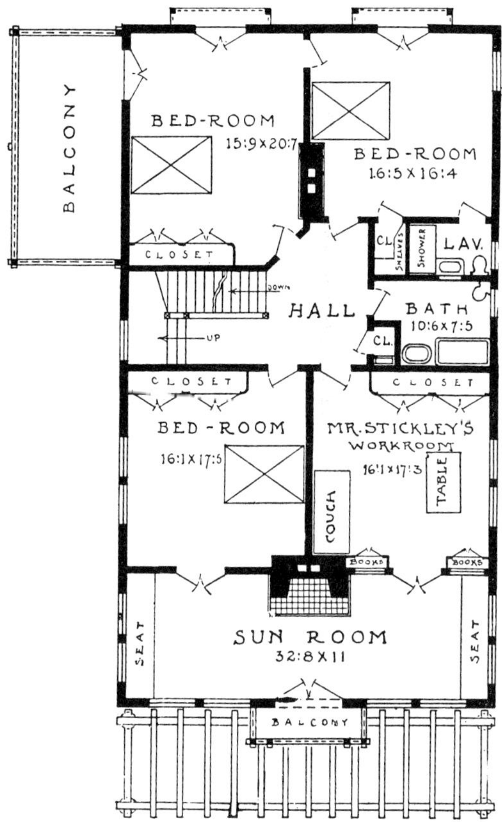

SECOND STORY FLOOR PLAN.

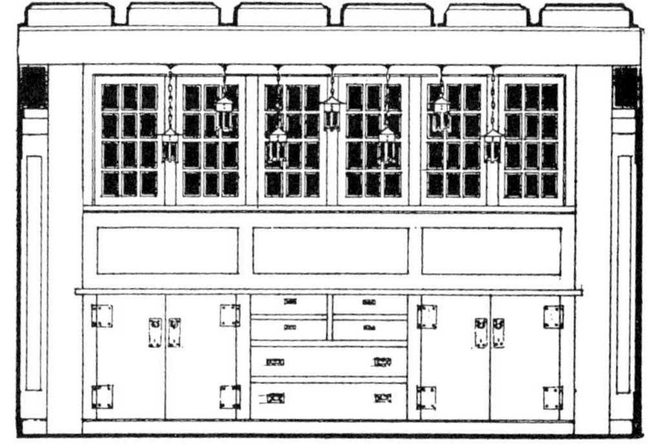

DETAIL OF DINING ROOM SHOWING
BUILT-IN SIDEBOARD AND WINDOWS.

proper, but is merely the expression of an individual fancy for an outdoor dining room and a sort of camp cooking place. At the end is built an outdoor fireplace and a big rough chimney. The detail of this fireplace, with its hobs, crane, and two brick ovens, is given in the first illustration.

The timbers are not applied to the outside of the house for the purpose of ornamentation, but are a part of the actual construction, which is thus frankly revealed. They are peeled chestnut logs squared on either side and with the face left rounded in the natural shape of the tree, hewn a little here and there to keep the lines from being exaggerated in their unevenness. These timbers are stained to a grayish brown tone that, from a little distance, gives the same effect as the bark. The lines of the red-tiled roof are low and broad, with an overhang of four feet on the ends and three feet at the sides.

The pergola is made of peeled cedar logs left in their natural shape and color, and the floor, which is almost on a level with the ground, is a dull red vitrified brick laid in herring-bone pattern at right angles. Extending from the side of the house is a roofed pergola,—if such a thing may be,—for while the timbers and the flooring are those of a pergola, it has a tiled roof like that of the house. This is not a part of the construction

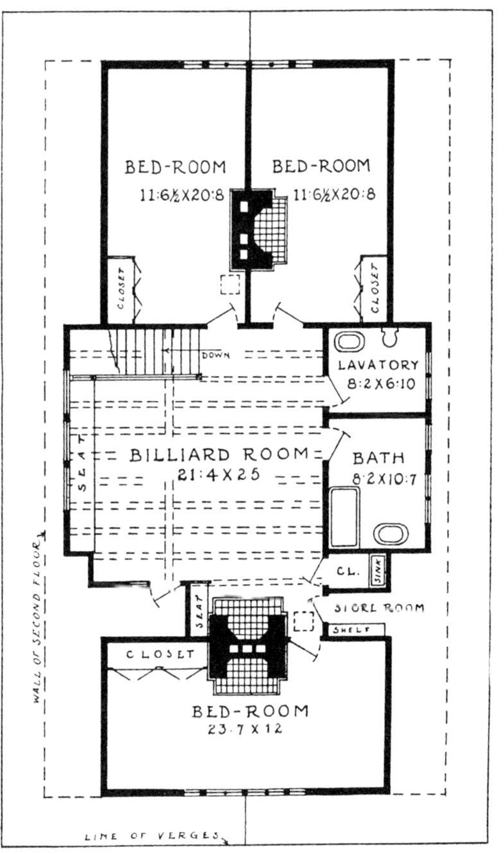

THIRD STORY FLOOR PLAN.

A SMALL SHINGLED HOUSE THAT SHOWS MANY INTERESTING STRUCTURAL FEATURES

Published in The Craftsman, February, 1907.

EXTERIOR VIEW FROM THE FRONT.

WE have suggested the use of shingles for the walls of this plain little cottage because they seem the best adapted to the peculiarities of its construction. They should, however, be laid in double course, the top ones being well exposed and the under ones showing not much over an inch below. This not only gives an interesting effect of irregularity as to the wall surface, but adds much to the warmth of the house. All the lines of the framework are simple to a degree, but the plainness is relieved by the widely overhanging eaves and rafters of the roof, the well-proportioned porch, which is balanced by the extension to the rear, the heavy beams which run entirely around the walls with a slight turn of the shingles above and the effective grouping of the windows. The little house is

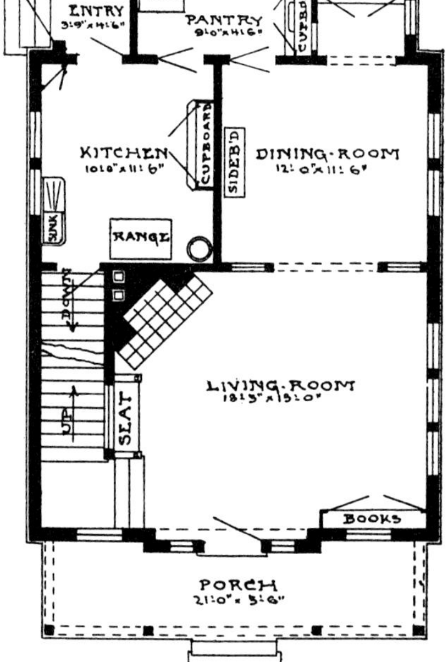

FIRST STORY FLOOR PLAN.

·WINDOW·SEAT·IN·LIVING·ROOM·

·INTERIOR·ELEVATION·OF·LIVING·ROOM·
·FACING·FRONT·OF·HOUSE·

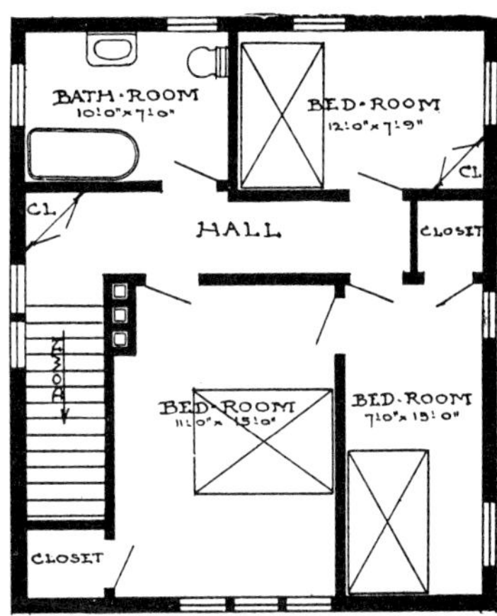

SECOND STORY FLOOR PLAN.

built to stand rough weather and this sturdiness is the direct cause of the wealth of attractive structural features. The roof of the porch projects two and a half feet, which affords protection even in a driving storm. Also for protection, all the exposed windows are capped by little shingled hoods which come up from the walls and which, in addition to their usefulness, form one of the most charming features of the whole construction. The eaves of the main roof project over the front for two and a half feet, and the weight is supported by purlins placed at the peak of the roof and at this connection with each of the side walls. This widely projecting roof gives a most comfortable effect of shelter and homelikeness, an effect which is heightened by the way in which the quaint little casement windows on the second story seem to hide under its wing. The view of the living room shown in the illustration is that which would be seen by anyone looking through the triple casement on the side wall. The first thing seen by one entering from the porch would be the fireplace, which is thrown diagonally across the corner with a small built-in seat between it and the landing of the staircase. The fireplace is made of rough red brick, with a stone mantel-shelf set on a line with the wainscot.

LIVING ROOM SHOWING CORNER FIREPLACE, BUILT-IN SEAT AND STAIR LANDING, WITH A VIEW OF THE ENTRANCE DOOR AT THE SIDE.

A SIMPLE, STRAIGHTFORWARD DESIGN FROM WHICH MANY HOMES HAVE BEEN BUILT

Published in The Craftsman, January, 1909.

EXTERIOR VIEW, SHOWING WELL-BALANCED PROPORTIONS AND SIMPLE TREATMENT OF WINDOWS AND WALL-SPACES.

FIREPLACE IN LIVING ROOM, SHOWING THE BUILT-IN BOOKCASES AND CASEMENTS ON EITHER SIDE.

A SIMPLE, STRAIGHTFORWARD DESIGN

THIS has been one of the most popular of the Craftsman house designs and as shown here it has been modified somewhat from the first plan, the modifications and improvements having been suggested by the different people who have built the house, so that they are all valuable as the outcome of practical experience. Although the illustration shows plastered walls and a foundation of field stone, the design lends itself quite as readily to walls of brick or stone, or even to shingles or clapboards, if a wooden house be desired.

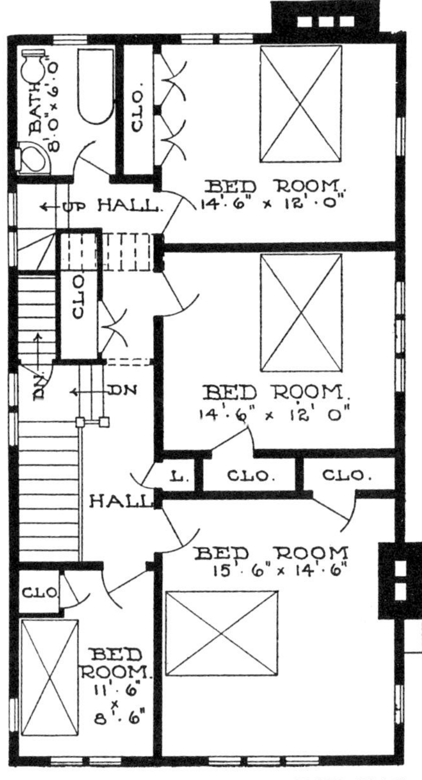

SECOND STORY FLOOR PLAN.

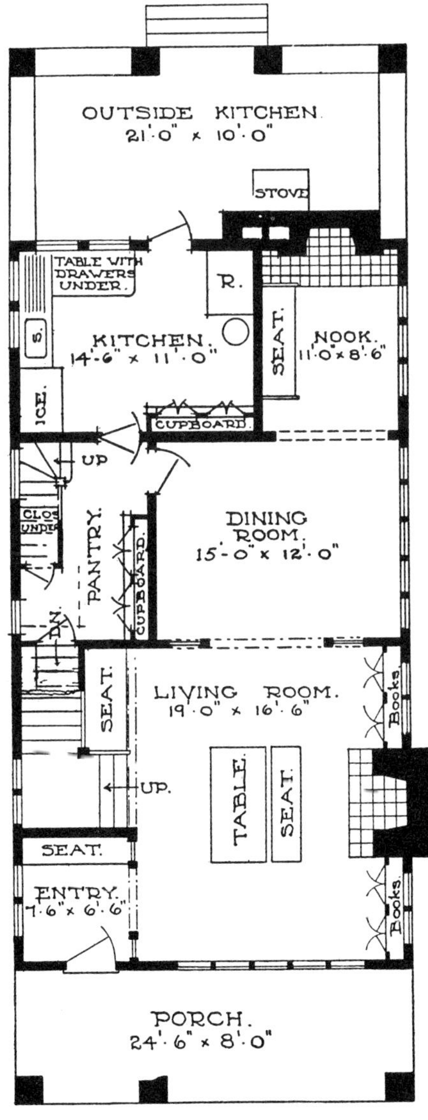

FIRST STORY FLOOR PLAN.

The outside kitchen at the back is recommended only in the event of the house being built in the country, because in town it would hardly be needed. In a farmhouse such an outside kitchen is most convenient as it affords an outdoor place for such work as washing and ironing, canning, preserving and other tasks which are much less wearisome if done in the open air. The position of the chimney at the back of the house makes it possible for a stove to be placed upon this porch for the use mentioned. The house is so designed that this outside kitchen may be added to it or omitted, as desired, without making any difference to the plan as a whole. The plan of the lower story shows the usual open arrangement of the Craftsman house. The entrance door opens into a small entry screened from the living room by heavy portiéres, so that no draught from the front door is felt inside. On the outside wall of the living room is the arrangement of fireplace and bookcases, as shown in the illustration. A large table might be placed in the center, with a settle back to it and facing the fire.

A CRAFTSMAN HOUSE IN WHICH TOWER CONSTRUCTION HAS BEEN EFFECTIVELY USED

FRONT VIEW OF HOUSE SHOWING TOWERS AND VERANDAS IN FRONT AND PERGOLA AT THE SIDE.

SOMETHING of a departure is made from the usual style of the Craftsman house in planning this one, which we regard as one of the most completely successful house plans ever published in THE CRAFTSMAN. It is not a large house, yet it gives the impression of dignity and spaciousness which usually belongs only to a large building; it is in no sense an elaborate house, yet it is decorative,—possessing a sort of homely picturesqueness which takes away all appearance of severity from the straight lines and massive walls. This is largely due to the square tower-like construction at the

A CRAFTSMAN HOUSE WITH TOWER CONSTRUCTION

All the exterior wood trim is of cypress very much darkened by the chemical process which we use. In this house the exterior woodwork is especially satisfying in its structural form, being decorative in its lines and the division of wall spaces and yet obviously an essential part of the structure. The horizontal beams serve to bind together the lines of the whole framework, and the uprights are simply corner-posts and continuations of the window frames. The roof of dull red tiles gives life and warmth to the color scheme of the exterior, and the thick round pillars painted white lend a sharp accent that emphasizes the whole.

The entrance door is at the left end of the porch which, by this device, is made to seem less like a mere entrance and more like a pleasant gathering place where outdoor life may go on.

FIRST STORY FLOOR PLAN.

two corners in front and to the upper and lower verandas, both ample in size and deeply recessed, which occupy the whole width of the house between the towers. Of these, one is the entrance porch and the other an outdoor sleeping room,—the latter a very essential part of every house that is built with special reference to health and freedom of living.

As suggested here, the house is of cement and half-timber construction with a tiled roof and a foundation of local field stone carefully split and fitted. The foundation is carried up to form the parapets that shelter the recessed porches on the lower story, and the copings are of gray sandstone. The walls are of cement plaster on metal lath, the plaster being given the rough gravel finish and colored in varying tones of green.

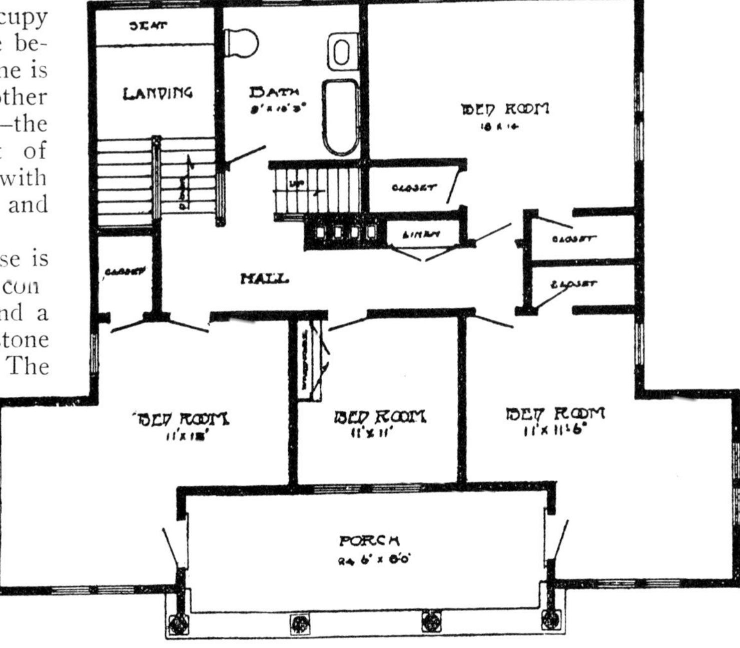

SECOND FLOOR PLAN. STORAGE-ROOM AND SERVANTS' ROOM IN ATTIC.

A CRAFTSMAN HOUSE WITH TOWER CONSTRUCTION

ALCOVE IN THE DINING ROOM MADE BY THE TOWER CONSTRUCTION. THIS LITTLE NOOK IS FITTED UP FOR A SMOKING ROOM OR DEN AND HAS ALL THE ADVANTAGES OF A BAY WINDOW OR SMALL SUN ROOM, AS THE WINDOWS ON EITHER SIDE ADMIT FLOODS OF SUNSHINE, PROVIDED THE HOUSE IS SO PLACED AS TO GIVE THIS TOWER A SOUTHERN OR WESTERN EXPOSURE. NOTE THE CONSTRUCTION OF THE OVERHEAD BEAMS.

A CRAFTSMAN HOUSE WITH TOWER CONSTRUCTION

A CORNER OF THE LIVING ROOM, LOOKING INTO THE DINING ROOM SO THAT THE POST-AND-PANEL CONSTRUCTION WHICH INDICATES A DIVISION BETWEEN THE TWO ROOMS IS PLAINLY SHOWN. THE CHIMNEYPIECE IS MADE OF LARGE SQUARE TILES, MATT-FINISHED IN A DULL TONE OF BROWNISH YELLOW AND BOUND AT THE CORNERS WITH STRIPS OF EITHER COPPER OR IRON. THE FIREPLACE HOOD IS OF COPPER AND THE ANDIRONS OF WROUGHT IRON. COMBINED WITH THE BROWN OF THE OAK OR CHESTNUT WOODWORK, THIS WOULD FORM THE BASIS OF A RICH AND QUIET COLOR SCHEME.

A BUNGALOW OF IRREGULAR FORM AND UNUSUALLY INTERESTING CONSTRUCTION

Published in *The Craftsman*, April, 1907

VIEW OF THE BUNGALOW SHOWING COURT AND PERGOLA, DINING PORCH AND SLOPE OF THE HILL.

THE plans and drawings of this bungalow, while partly our own, are adapted from rough sketches sent us by one of our subscribers, Mr. George D. Rand, of Auburndale, Mass. Mr. Rand is an architect who has retired from active work, and these sketches were made for his own bungalow, which is situated in the mountain region of New Hampshire. In sending us the sketches, Mr. Rand kindly gave us permission to use the idea as outlined by him, with such alterations as seemed best to us. In accordance with this permission, we make quite a number of minor modifications in the original design, and many of the suggestions for construction are our own.

The house is somewhat irregular in design, but is so admirably proportioned and planned that the broken lines impress one as they do when seen in some old English house that has grown into its present shape through centuries of alteration in response to changing needs. It seems above all things to be a house fitted to crown a hilltop in the open country, especially where the slope is something the same as indicated in the site here shown. The line from the back of the roof down to the boat landing comes as near to being a perfect relation of house and ground as is often seen, and this relation is of the first importance in the attempt to suit a house to its environment.

The exterior walls and the roof are of shingles, and the foundations, parapets, columns and chimneys are of split stone laid up in dark cement. The construction of the roof is admirable and, with all the irregularity, there is a certain ample graciousness and dignity in line and proportion. At the front of the house between the two gables is a recessed court, paved with red cement cut into squares like tiles and roofed over with a pergola of which the beautiful construction is shown in the detail given of this court.

The large porch at the side of the house is intended for an outdoor living and dining room and corresponds closely in arrangement with the rooms which open upon it. Its construction is the same as that of the court, except that it is sheltered by a wide-eaved roof instead of a pergola and is so arranged that it can be easily closed in for cold or stormy weather. At the end next the living room there is a large fireplace built of split stone, which exactly corresponds with the fireplace in the indoor living room. A good fire of logs on this outdoor hearth gives the same effect of warmth and cheer as a camp fire. If casements were placed all around the porch so that it could be entirely closed in time of storm and cold, it might be an excellent idea to floor it smoothly with wood for dancing; but if

it is to be exposed to the weather, the cement floor would be more durable, as sun and wind soon roughen the best wood floor.

The house is rich in fireplaces, for not only are there the large chimneypieces, in the living room and on the porch adjoining, but two of the bedrooms on the lower floor have corner fireplaces. As the kitchen is so placed as to be practically detached from the remainder of the house, another flue is necessary for the kitchen range.

From the court the entrance door opens into a small square hall, which is practically an alcove of the living room and which connects by a narrow passage with the bedrooms at the opposite side of the house. The bathroom is placed almost in the center of the house, which might be undesirable if it were not completely shut off from the living rooms by the plan of the hall and by the same plan made easily accessible to the three bedrooms.

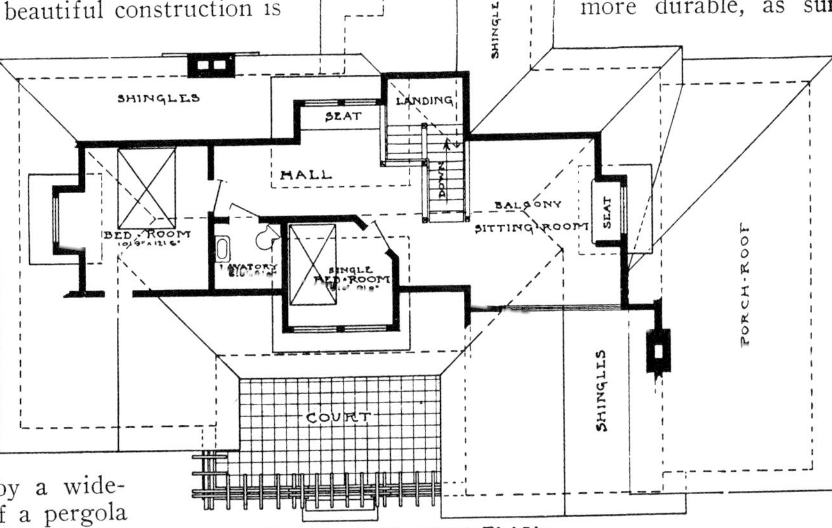

DETAIL OF THE COURT, SHOWING CONSTRUCTION OF THE PERGOLA AND THE USE OF VINES, SHRUBS AND FLOWER BOXES. THE GLASS DOOR IS THE MAIN ENTRANCE DOOR OF THE HOUSE.

The construction of the living room is very interesting, as everything is revealed up to the ridge pole and rafters of the roof. The roof itself has such a long sweep that there would be danger of its sagging, were it not for the trusses that brace it in the center. These trusses, in addition to their use, add much to the decorative effect of the structure. Across the front and down the side of the living room to the fireplace is a built-in seat paneled below and backed with a wainscot of V-jointed boards. If desired, the top of this seat can be hinged in sections, making the lower part a place for storing things. The window above the seat in front gives an unusually interesting effect, as there is a group of double casements on what in an ordinary house would be the lower floor, and another group of single casements, the center one higher than the sides, just above the frieze and beam. Another casement set high in the wall is placed opposite the fireplace, corresponding in position to the door which opens upon the porch.

Extending to a point half way across the opening into the hall is the balcony which forms the upstairs sitting room; this is divided from the living room only by a railing. The floor of this balcony forms the ceiling of the dining room, which is separated from the living room only by double cupboards made to be used as bookcases on one side and china closets on the other. These cupboards extend to the same height as the window-sills and mantel, carrying this line around the room. The space above is open and hung with small curtains. This effect of a small low dining room recessed from a living room that runs clear to the roof is delightful in the sense it gives of homelike comfort, as the effect is that of a snug little retreat devoted to good cheer.

CORNER OF THE LIVING ROOM, SHOWING UPPER AND LOWER WINDOWS AT THE FRONT OF THE HOUSE, THE SEAT WHICH SERVES BOTH AS WINDOW AND FIRESIDE SEAT AND THE FORM AND CONSTRUCTION OF THE FIREPLACE.

A PORTION OF THE LIVING ROOM, LOOKING INTO THE DINING ROOM. THE CEILING OF THE LATTER IS FORMED BY THE FLOOR OF THE BALCONY ABOVE, SO THAT IT HAS THE APPEARANCE OF A LOW-CEILED RECESS, AND THE BOOKCASES MAKE THE PARTITION. THE BALCONY IS USED AS AN UPSTAIRS SITTING ROOM.

TWO INEXPENSIVE BUT CHARMING COTTAGES FOR WOMEN WHO WANT THEIR OWN HOMES

IT has always seemed to us that if there is one kind of dwelling that is more generally needed than another, it is the small and inexpensive, yet comfortable and homelike, cottage that can be built almost for the year's rent of a flat, or even of room and board in a boarding house, and that would serve as a home for two or three people. Especially is this sort of a house needed by women of limited means,—women who either work at home or possibly in an office or shop and who need all the home comfort they can get, instead of dragging out an existence in a boarding house or facing the bugbear of rent day in a flat.

These cottages each would serve to accommodate a group of three or four and the number might even be stretched to six in case of very congenial people who did not mind sharing their rooms. The houses as represented here are built of field stone, but the designs would serve equally well for concrete, —a form of construction that would greatly lessen the cost,—or for frame houses covered with shingles, clapboards, or even with plain boards and battens. In fact, after the initial cost of the lot in some suburb not too far away from the place of employment, it should be a very easy matter for two or three women who felt that they would like to make a home for themselves to combine their resources and build one of these little houses. Even the cost of the lot might be very greatly lessened if it were possible to build in a village near the city or right out in the country. It is the woman who is stranded in some forlorn hall bedroom, or who is forced to feel that she is a superfluous member of someone else's family, who would most welcome the dignity and content that would be found

Published in The Craftsman, March, 1904.

STONE COTTAGE WITH RECESSED PORCH AND BUNGALOW ROOF.

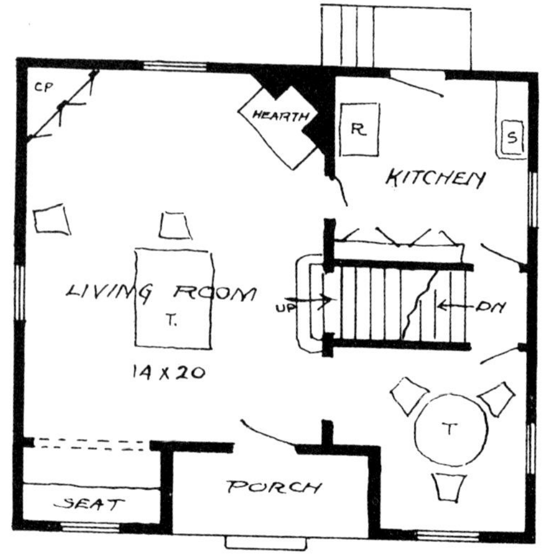

FIRST STORY FLOOR PLAN.

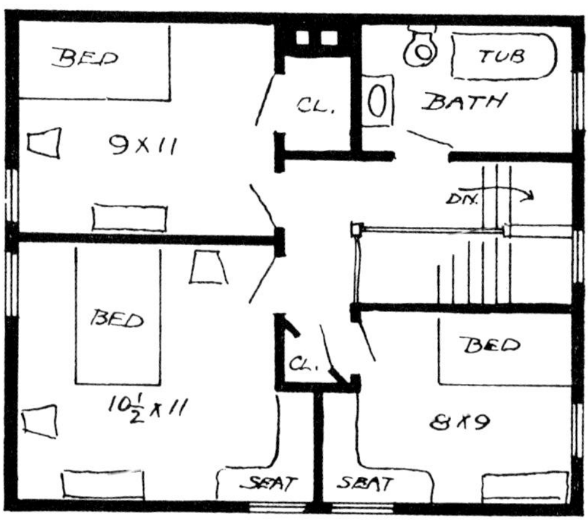

SECOND STORY FLOOR PLAN.

TWO INEXPENSIVE BUT CHARMING COTTAGES

in a home of her own,—a home which might be shared by a relative or close friend in similar circumstances.

The chief value of these little houses lies in the fact that although they are but the simplest of cottages, they nevertheless possess a beauty and individuality which is lacking in many a residence that costs ten times as much. We feel that in exterior attractions they are fitted to take rank with any of the houses designed in The Craftsman Workshops, and that the interior arrangement is compact and comfortable to a degree. The chief difference between them, as regards the exterior, lies in the fact that in the case of the first one the porch is recessed and, in the second, is extended to the dimensions of a good-sized veranda that runs the whole width of the house. In interior arrangement they are much alike, the living room in each case occupying the whole of one side of the house and opening into a dining alcove which takes about half of the other side. The kitchen occupies the remaining corner and, if this be fitted with convenient cupboards, work table and the like, there would be no necessity for a pantry. Upstairs also the arrangement of the two cottages is somewhat similar, as in each case the space is divided into three bedrooms and a bathroom, with plenty of closet room tucked away into nooks and corners.

As to the interior woodwork and furnishing, these need not be costly in order to be attractive. Some inexpensive native wood, such as pine, or cypress, or that grade of chestnut known to builders as "sound wormy," would, if finished properly, give the most delightful effect when used for interior trim, built-in seats, cupboards, balustrades for the stairways, and for wainscoting, —providing the sum set aside for the house admitted such a luxury as the last. The remaining wall spaces and the ceilings could be left in the rough sand-finished plaster, tinted in any color desired, and the fireplace would naturally be of brick or field stone and of the simplest design. Given such a foundation, the question of furnishing would adjust itself.

Published in The Craftsman, March, 1904.

STONE COTTAGE WITH VERANDA. NOTE THE EFFECT OF SQUARE BUNGALOW ROOF AND OF CASEMENT WINDOWS HIGH UNDER THE EAVES.

FIRST STORY FLOOR PLAN.

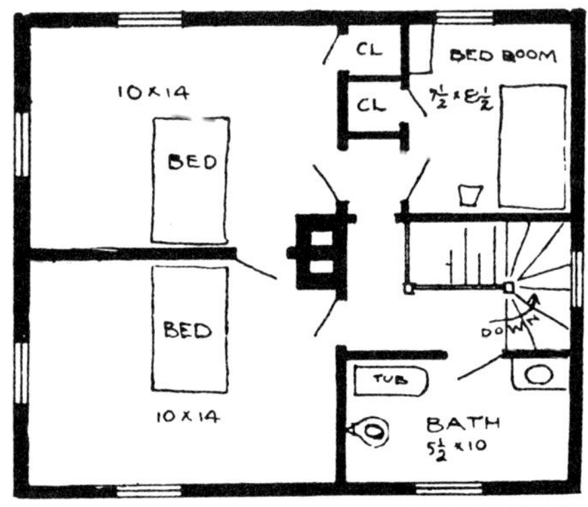

SECOND STORY FLOOR PLAN.

A LOG HOUSE THAT WILL SERVE EITHER AS A SUMMER CAMP OR A COUNTRY HOME

Published in The Craftsman, March, 1907.

EXTERIOR OF LOG HOUSE, SHOWING DECORATIVE USE OF THE PROJECTING ENDS OF PARTITION LOGS.

SO many people like log houses for summer homes that we give here a design that would harmonize with the most primitive surroundings. At the same time it is so carefully planned and so well constructed that it could be used as a regular dwelling all the year round. While the lines of the building are simple to a degree, all the proportions are so calculated and the details of the construction so carefully observed that, with all this simplicity and freedom from pretense, there is no suggestion of bareness or crudity. It is essentially a log house for woodland life, and it looks just that; yet it is a warm, comfortable, roomy building perfectly drained and ventilated and, with proper construction, ought to last for many generations.

As the first step towards securing good drainage and also saving the lower logs of the wall from decay, there is an excellent foundation built of stone or cement,—according to the material most easily and economically obtained,—and this foundation is quite as high as it would be in any dwelling built of the conventional materials in the conventional way. But as the appearance of such a foundation would spoil the whole effect of the house by separating it from the ground on which it stands, it is almost entirely concealed by terracing the soil up to the top of it and therefore to the level of the porch floors. The first log of the walls rest directly upon this foundation and is just far enough above the ground to prevent rotting. By this device perfect healthfulness is secured so far as good drainage is concerned, and at the same time the wide low house of logs appears to rest upon the ground in the most primitive way.

The logs used in building should have the bark stripped off and then be stained to a dull grayish brown that approaches as closely as possible to the color of the bark that has been removed. This does away entirely with the danger of rotting, which is unavoidable when the bark is left on, and the stain removes the raw, glaring whiteness of the peeled logs and restores them to a color that harmonizes with their surroundings. The best logs for this purpose are from trees of the second growth, which are easily obtained almost anywhere. They should be from nine to twelve inches in diameter and should be carefully selected for their straightness and symmetry.

The wide porches that extend all along both sides of the house afford plenty of room for outdoor living. As shown in the picture, one end of the porch at the front of the house is recessed to form a square dining porch, which opens into the kitchen and also into the big room. This is a combined living room and indoor dining room, to be used for the latter purpose only in chilly or stormy weather, if the house is meant for a summer camp.

The general effect of this room is in exact harmony with the exterior of the house. The door from the porch opens into an entry which on one side gives access to the two bedrooms at the front of the house and on the other leads by a wide opening into the main room. The walls and partitions are of logs and the ceiling is beamed with logs flattened on the upper side to support the floor above. The fireplace, like the chimney outside, is built of split stone, a material especially suited to this house, and is in a nook or recess that is formed, not by the shape of the room, but by the suggestion of a division made by the two logs placed one above the other across the ceiling logs, and the two posts that form the ends of the fireside seats.

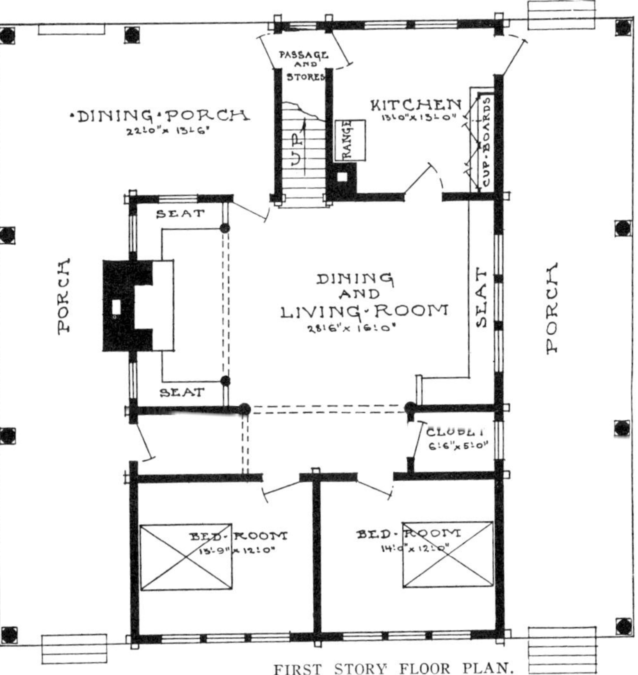

FIRST STORY FLOOR PLAN.

VIEW OF LIVING ROOM, SHOWING THE LOG CONSTRUCTION WHICH SEPARATES THE FIREPLACE NOOK FROM THE REST OF THE ROOM, AND ALSO GIVING AN IDEA OF THE EFFECT TO BE OBTAINED BY THE USE OF LOG PARTITIONS.

ARCHITECTURAL DEVELOPMENT OF THE LOG CABIN IN AMERICA

WHAT is there about a log cabin that appeals to our imagination, that seems so alluring and full of the suggestion of romance? Is it not because the house of logs is a part of our heredity? It was a primitive home to man, a rudimentary sheltering of domestic life, a place of safety where love and friendship could be shut in and foe and danger shut out. The early homes of our Germanic ancestors were huts in the forest, sometimes built around a central tree which grew up through the roof and spread its sheltering branches over the dwelling. We came from the forest, and trees formed our home and our protection. And so today a house built of wood which has not been metamorphosed into board and shingle, but still bears the semblance of the tree, rouses in us the old instinctive feeling of kinship with the elemental world that is a natural heritage.

To us in America the log cabin seems a near friend. For many of us it was the home of our immediate ancestors, and it forms a vital part of the life of the white man in this continent. What a train of historical reminiscence the mere thought of the log cabin awakens: the landing of the first settlers, the unbroken wilderness of the primeval forests, the clearing of the ground, the building of the first homes. How great must have been the need of the comfort of the hearth and the strength of fellowship in that lonely and desperate struggle against the elements, the foe and starvation! Scattered far over this continent, moving northward, southward and westward, the log cabin has been the pioneer of civilization, the sign of the determination of the white man to face the unknown and to conquer all obstacles. Viewed in this light, it seems of a certain poetic significance that Lincoln, one of the greatest of the nation's leaders, should have been born and reared in a log cabin.

Since the log house has played so important a part in our history its development into a definite and characteristic type of architecture might give us something national, something peculiarly American in suggestiveness. There are elements of intrinsic beauty in the simplicity of a house built on the log cabin idea. First, there is the bare beauty of the logs themselves with their long lines and firm curves. Then there is the open charm of the structural features which are not hidden under plaster and ornament, but are clearly revealed—a charm felt in Japanese architecture, which is, as Cram has said, "the perfect style in wood as Gothic is the perfect style in stone." The Japanese principle, "The wood shall be unadorned to show how beautiful is that of which the house is made," is true of the Craftsman development of the log house. For in most of our modern houses "ornament by its very prodigality becomes cheap and tawdry," and by contrast the quiet rhythmic monotone of the wall of logs fills one with the rustic peace of a secluded nook in the woods.

Of the distinction and charm of such a type the log house at Craftsman Farms is a proof, for it is a log cabin idealized. Some idea of its homelike beauty can be gathered from the views and floor plans given here. And in addition several smaller and simpler forms of this construction are shown—little log bungalows for summer or all-year use, in woods, or mountains, or by the shore.

THE LOG HOUSE BUILT AT CRAFTSMAN FARMS

VIEW OF LOG HOUSE AT CRAFTSMAN FARMS, SHOWING ENCLOSED PORCH, FIFTY-TWO BY FOURTEEN FEET.

A S in the pioneer days, the space for this house had first to be cleared in the forest. The abundant chestnut trees were cut down and of them the house is built. The logs are hewn on two sides and peeled and the hewn sides laid together and chinked with cement mortar. The logs are stained the color of the bark.

A stone foundation runs under the whole building, including the wide veranda across the front of the house. The most practical piece of furniture here is a long combination bench and wood-box in which is kept the smaller wood for the fires.

From the veranda a wide door leads into the great living room with its fireplace at either end. The large hearths, which have special ventilating and heating appliances, are built of field stone gathered on the place and are topped with low-hanging hoods. Most of the available wall space is filled by bookcases. Above the bookcases and over the settles are windows with small diamond panes, and the light is softened to a mellow glow by casement curtains of burnt orange. The color scheme of the whole room reminds one of the forest— brown and green, with the glint of sunshine through the leaves, suggested by the gold of the windows and the gleam of copper in the

hearth-hoods, the door-latches and the vases and bowls on the bookcases and table.

The dining room runs parallel to the living room. Here also is a big ventilating hearth. These fireplaces heat the entire house with hot water and warm air.

Beyond the dining room is the kitchen, a large room, light and airy, painted white, with a large range. There are special appliances for convenience in washing dishes, etc.

The main bedrooms on the second floor are at the two ends of the house; one of them is furnished and decorated in yellow and seems aglow with sunshine; the other, a much larger room, is done in blue and gray with woodwork of dark gumwood. The walls are covered with gray Japanese grass-cloth, the hearth is of dull blue Grueby tiles with a brass hood, and the furniture is gray oak.

The illustrations give but an inadequate idea of the charm and comfort of the interior, its harmony with nature and its unity of the best of civilization with the best in cruder forms of life. Three views are given of the great living room—the first showing the big stone fireplace at one end with bookshelves at the side and Craftsman chairs and table grouped around the hearth. A second view, at the top of page 149, gives one some impression of the

SIDE VIEW OF LOG HOUSE AT CRAFTSMAN FARMS.

DETAIL IN ONE END OF THE LIVING ROOM, SHOWING CRAFTSMAN FURNITURE ABOUT THE BIG STONE FIREPLACE.

VIEW OF GREAT LIVING ROOM, WITH STAIRCASE AT THE LEFT AND STONE CHIMNEY AT THE FURTHER END.

DETAIL OF LIVING ROOM, WITH ENTRANCE INTO THE DINING ROOM: THE LOG CONSTRUCTION IS INTEREST-INGLY SHOWN HERE.

ONE END OF THE DINING ROOM IN THE LOG HOUSE: THE FURNITURE IS FROM THE CRAFTSMAN WORKSHOPS.

SIDE OF DINING ROOM, SHOWING LONG CRAFTSMAN SIDEBOARD: THE LOGS IN THIS ROOM ARE FINISHED WITH A WOOD OIL WHICH GIVES A DELIGHTFUL MELLOW TONE AS THOUGH SUNLIGHT WERE POURING INTO THE ROOM.

ONE CORNER OF THE LARGE BEDROOM ON THE SECOND FLOOR OF THE LOG HOUSE, WITH ALCOVE FOR BED.

wide hospitable spaces of the interior, showing the long vista down the room with the other fireplace at the further end, the staircase on the left and the big glass doors opening onto the veranda at the right. The lower view on the same page shows a detail of the room, including the long table, the piano, and a glimpse through the curtained opening into the dining room beyond. Here again one feels the harmony of the carefully-designed furnishings with the more primitive dignity of the log construction. The remaining views show dining room and one of the bedrooms.

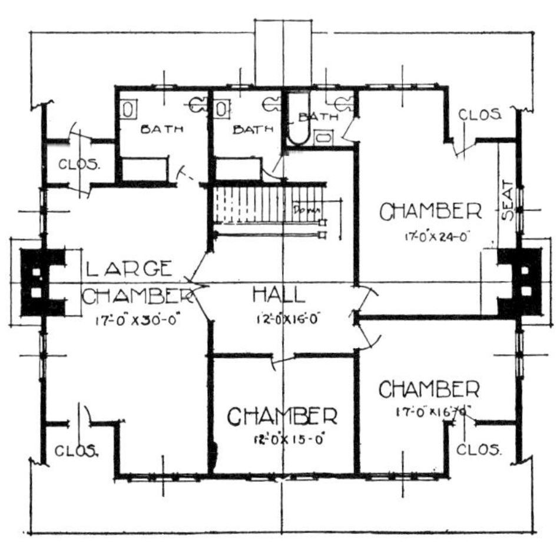

THE LOG HOUSE: FIRST FLOOR PLAN.

THE LOG HOUSE: SECOND FLOOR PLAN.

LOG COTTAGE FOR SUMMER CAMP OR PERMANENT COUNTRY HOME

Published in The Craftsman, March, 1907.

CRAFTSMAN COTTAGE OF LOGS, WITH STONE FOUNDATION AND CHIMNEY, AND SHINGLED ROOF: THE TWO LONG PORCHES AND THE DINING PORCH AT THE REAR GIVE AMPLE ROOM FOR OUTDOOR LIVING: NO. 48.

INTERIOR VIEW IN COTTAGE NO. 48, SHOWING LIVING-ROOM INGLENOOK WITH STONE CHIMNEYPIECE AND BUILT-IN CORNER SEATS, AND DINING TABLE IN THE FOREGROUND: THE NATURAL USE OF THE LOGS AND DECORATIVE EFFECT RESULTING FROM PRACTICAL HANDLING OF STRUCTURAL FEATURES IS MOST HOMELIKE.

LOG COTTAGE FOR SUMMER CAMP OR PERMANENT HOME

ALTHOUGH this log bungalow is primarily intended for a summer home, it is so carefully planned and so well constructed that it could be used as a regular dwelling all the year round. While the lines of the building are simple to a degree, the proportions and details have all been so thoughtfully considered that with all this simplicity and freedom from pretense there is no suggestion of bareness or crudity. It is essentially a log cabin for woodland life, and looks just that; yet it is a warm, comfortable, roomy building, perfectly drained and ventilated, and if properly built ought to last for many generations.

There is a foundation of stone or cement, sufficiently high to secure good drainage and save the lower logs from decay. This foundation, however, is almost entirely concealed by terracing the soil up to the top of it, to the level of the porch floors. By this means perfect healthfulness is secured, and at the same time the wide, low cottage of logs appears to rest upon the ground in the most primitive way.

The logs used in building have the bark stripped off and are stained to a dull grayish brown that approaches as closely as possible the color of the bark. The removal of the bark prevents rotting, and the stain restores a color that harmonizes with the surroundings.

The wide porches afford plenty of room for outdoor living. One porch is recessed at the end to form a square dining porch, which opens into the kitchen and also into the big room which is a combined living room and indoor dining room.

The entry opening from the porch gives access on one side to the two bedrooms, and on the other leads by a wide opening into the main room. The walls and partitions are of logs and the ceiling is beamed with logs flattened on the upper side to support the floor above. The fireplace, like the chimney outside, is built of split stone, and is in a nook formed or suggested by the two logs placed one above the other across the ceiling logs, and by the two posts at the ends of the fireside seats.

The perspective view of the living room shows what a decorative effect results from this simple rustic treatment, and how entirely the furnish-

ings are in keeping with the purpose and character of the log construction. There is a primitive, picturesque quality about the whole that would be lacking in a more formal interior, and the natural use of the logs seems to relate the rooms very closely to the exterior of the cottage and the woods and hills around.

The upper story may be arranged as the builder pleases. If intended for a permanent home, it can be divided into bedrooms and a bath, but for camp life in the woods a large single room may be left where things can be stored and cots put up or hammocks slung.

The expense of furnishing the bungalow would be somewhat reduced by the built-in fireside seats in the living room and the long seat which stretches beneath the windows in the opposite wall. These could be made with hinged tops, thus providing very useful storage space in addition to the large closet between the living room and bedroom, the kitchen cupboards and the space below the stairs.

If a larger bedroom were desired on the first floor, the entry shown in the plan might be omitted, including this space within the adjacent bedroom and using the door in the corner of the recessed dining porch as the entrance door. Or if only one bedroom were needed a room might be used as library.

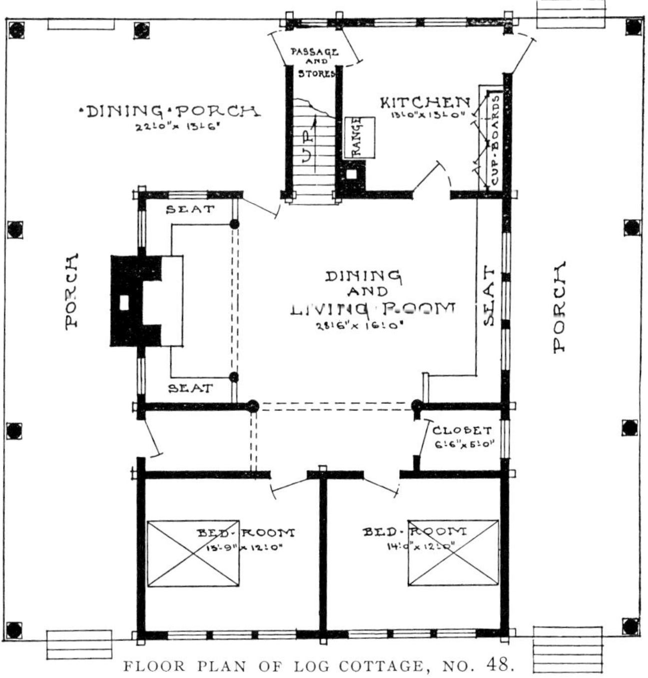

FLOOR PLAN OF LOG COTTAGE, NO. 48.

SMALL BUT COMFORTABLE LOG DWELLING

Published in The Craftsman, March, 1907.

LOG HOUSE WITH RUSTIC INTERIOR: NO. 49.

A HILLTOP or slight elevation is the site for which this little cottage was designed, and the ground is terraced to the level of the porch floors, concealing the foundation and seeming to connect the building more closely with the soil. The main roof and that of the porch are shingled, and the columns supporting the latter are thick peeled tree trunks, harmonizing with the peeled and stained logs of the walls. The chimneys are made of split stone. Flower-boxes at the upper windows give a little touch of grace and color that is

LIVING ROOM IN LOG HOUSE NO. 49, WITH UNIQUE CHIMNEYPIECE CONSTRUCTION AND RUSTIC FIRESIDE SEATS

unusually attractive against the brown background of logs. The porch is floored with red cement and the steps leading up to the house are of split stone laid in cement and smoothed off.

The central living room is simple in construction, but there is a dignity in the unbroken lines of the logs that is very effective. An interesting structural feature of the stone chimney-piece is the framing made by the ends of the logs forming the partition on each side of the bathroom, and the log that crosses the mantel-breast at the top, like a lintel. On either side of this fireplace is a large settle built of peeled saplings stained to the same color as the logs of the house. The supports for the seat cushions and the backs are made of ropes twisted and knotted around the frame of the settle. The seat cushions and pillows are of canvas or some such sturdy material in keeping with the rustic interior.

There are two downstairs sleeping rooms, one on each side of the bath. The upper story, which is left undivided, has plenty of light and ventilation, so that it could easily be partitioned into rooms if desired. Casement windows are used throughout, hooded where exposed to the weather. Dutch doors, V-jointed,

RUSTIC PERGOLA FITTED UP FOR OUTDOOR LIVING.

and with large strap hinges, are used for outside doors.

In this sort of cottage, where the porch is likely to be much in use for outdoor meals and as a cool, shady place for work, rest or play, rustic furniture would be as practical as it would be charming. A long seat, a few chairs and a table would serve all the purposes of usefulness and comfort, and would be in perfect harmony with the rest of the building, helping to carry out the rustic effect of the log construction. Or perhaps down by the water's edge or in some other pleasant spot a rustic pergola could be erected, like the one shown above, with vine-clad pillars and roof, inviting chairs, a little table for books or sewing or afternoon tea, and possibly a swinging seat suspended between two of the posts. Thus the hospitality of the log dwelling would be extended as far as possible into the nature world around it, luring one always from the shelter of the cottage into the fresh mountain air, while the unity of the log construction would link the little home more closely to the surrounding woods of which its walls were once a living part. In fact there are various possibilities for practical and picturesque constructions of this sort which ingenuity and skill could devise, at little labor and expense.

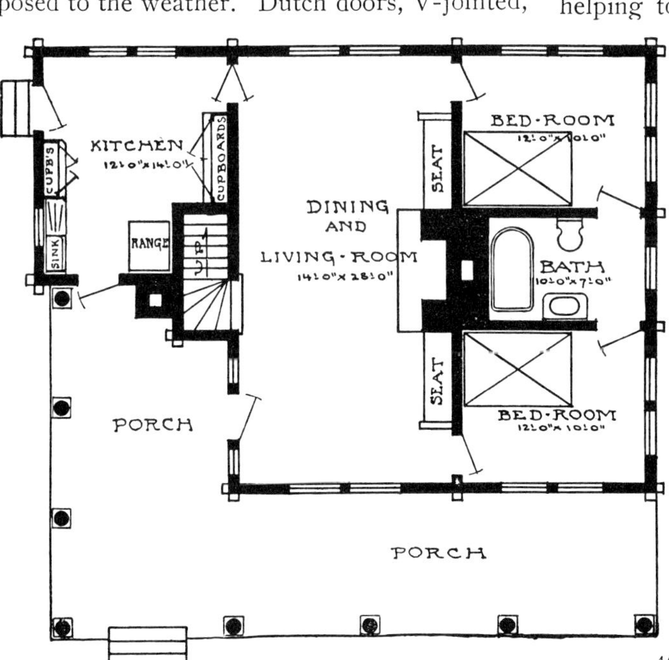

FLOOR PLAN OF LOG HOUSE NO. 49

97

LITTLE WOOD COTTAGE ARRANGED FOR SIMPLE COUNTRY LIVING

Published in The Craftsman, March, 1907.

RUSTIC COTTAGE BUILT OF SLABS: NO. 50.

THIS cottage is built of slabs, and while not actually as massive as log construction, gives the same effect of primitive and rugged comfort. The slabs are peeled, nailed to the sheathing of the walls, and stained as nearly as possible to the color of bark. The proportions of the building are low and broad. The widely spreading roof is shingled, the porch columns are logs peeled and stained, and the foundation is concealed under the terrace. The front porch serves all the purposes of an outdoor living room, and the one in the rear is intended for a dining porch, whenever the weather permits.

In the living room every feature of the construction is frankly revealed, and this forms the chief element of decoration. The deep nook that divides the porches is the center of comfort and restfulness for the whole house. Bookshelves are built in on each side of the big fireplace of split stone, and there are two large box-seats made of peeled slabs. The nook has no ceiling, but extends up into a gable, separated by a railing from the attic.

Another fireplace is built in the rear wall of the living room, in this instance the placing of seats being left to the choice of the owner. This end of the room, being between the dining porch and the kitchen, is intended to be used as an indoor dining room whenever the weather is not mild enough for meals to be carried out onto the porch. The front wall of the living room is filled by a group of three windows, and on the right are doors leading to the two bedrooms, in one of which is a long window seat. The kitchen is provided with ample shelf and storage room, and between the kitchen and bedroom walls the stairs lead up from the living room to the attic. This may be arranged as desired, either divided into several small bedrooms or left in one large room for storage purposes or to

SUGGESTION FOR A RUSTIC GATEWAY AND SEAT.

RUSTIC COTTAGE ARRANGED FOR SIMPLE COUNTRY LIVING

INTERIOR VIEW OF COTTAGE NO. 50, SHOWING SPACIOUS LIVING ROOM AND RECESSED FIREPLACE NOOK.

accommodate cots and hammocks when extra sleeping room is required.

The view of the living room and inglenook shown above suggests a satisfactory method of handling the structural features of the interior, the natural use of the woodwork being especially appropriate for this type of building. In fact, the arrangement of the bungalow, and the way in which the structure itself is made the basis of all decorative effect, are both an illustration and suggestion of how much can be accomplished by working along these practical, straightforward lines. While in this instance, of course, the design and scale of the cottage is of the most unpretentious, it embodies in its simple construction many of these characteristics which are typical of the Craftsman ideal home.

Rustic furniture for the porch would be serviceable and in keeping with the slab walls of the building, and the sketch of the rustic gateway, with its pergola roof draped with wistaria and the primitive charm of the little seat below, suggests an appropriate entrance to the garden, and may serve to suggest other variations on the same theme. A circular rustic seat, for instance, might be built around one of the neighboring trees.

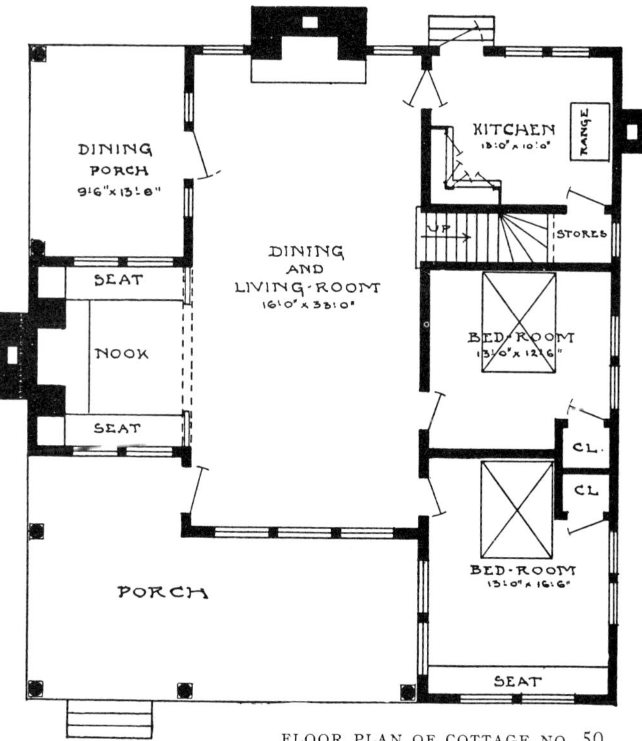

FLOOR PLAN OF COTTAGE NO. 50.

COMFORTABLE ONE-STORY BUNGALOW OF LOGS

ROUGH stone foundation, walls of log, and roof of colored Ruberoid are the materials used for this bungalow. The cement floor of the recessed entrance porch is extended beyond the house line as shown, and a log pergola is set on a cement floor at the rear of the building.

The floor plan has been worked out with an idea of economy of effort in housekeeping.

The entrance door opens directly from the porch into the living room, the deep recess other has a comfortable built-in seat. The open bookshelves on the other side of the door to the pergola and the groups of high casements break up the wide expanse of wall space and form an interesting group of furnishings.

Both bedrooms are separated from the living room by a narrow hall, and the bath is located between the bedrooms—an arrangement combining convenience and privacy. Ample closet room is provided, and the groups of windows

Published in The Craftsman, May, 1911.

FIVE-ROOM LOG BUNGALOW: NO. 115.

sheltering it sufficiently to make a vestibule unnecessary. On either side of the door are two casement windows with small panes. In fact this style of window is used throughout the building, being more suitable for a rural dwelling than the double-hung type, and the small panes breaking up the surface of the walls and adding considerably to the interest of both the exterior of the bungalow and the rooms within.

The center of comfort and charm of the interior is of course the open fireplace which occupies the center of one of the living-room walls and uses the same chimney as the flue of the kitchen range just behind.

A coat closet is conveniently located on one side of the fireplace, while the space on the are so situated as to allow cross ventilation.

On the opposite side of the house is the dining room, which, owing to the limitations of the floor plan, is separated from the main room more than is usual in a Craftsman interior. In the present instance, however, this arrangement may be found preferable, as it affords more opportunity for privacy and gives ready access between kitchen, pantry and dining room. Both the kitchen and pantry are arranged with the utmost convenience, with plenty of table and closet room and two sinks and although neither room is large the very compactness of the arrangement makes for economy of housework. The dining room, like the bedroom at the opposite corner, has front and side window groups as well as smaller

FIREPLACE CORNER OF LIVING ROOM IN BUNGALOW NO. 115, SHOWING CHIMNEYPIECE OF SPLIT FIELD STONE WITH RECESSED SHELF, AND INTERESTING USE OF WOOD FOR WALLS AND BUILT-IN FIRESIDE SEAT

casements placed rather high in the wall overlooking the front recessed porch.

The little rear porch built under the main roof adjoining the kitchen may be glassed in winter and screened in summer, and will thus serve as an additional room for kitchen and laundry work.

The interior view given above shows the fireplace corner of the living room and gives a general idea of the appearance of the rest of the interior. The chimneypiece of split field stone, with the deeply recessed shelf, though simple in construction forms a very effective part of the structure, and emphasizes the hospitable air of the spacious room. The corner seat on the left, with its wainscoted back of wide V-jointed boards, could be made very comfortable with a few pillows, and the casement windows in the wall above as well as the small glass panes of the door that leads to the pergola porch at the rear, all help to make the room a pleasant, homelike place.

In this bungalow, as in most of the others shown, a few pieces of rustic furniture grouped about the porch would help to increase its

hospitality and comfort, and would insure the bringing of many little household tasks into the fresh air whenever the weather permitted. Seats could be placed on each side of the front porch beneath the casement windows, and rustic fittings could be used for the rear porch.

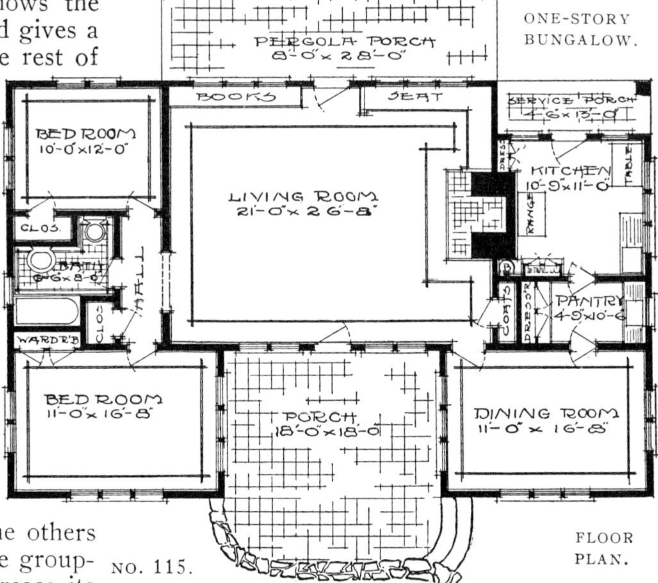

101

PERMANENT SUMMER CAMP OF LOGS WITH TOP STORY ARRANGED FOR OUTDOOR SLEEPING

Published in The Craftsman, August, 1911.　　　　　PERMANENT SUMMER LOG CAMP: NO. 121.

WE have planned here a log building for a summer camp, so inexpensively and simply built that it can be closed with safety during the winter and easily put into livable order each spring.

The logs are placed upright and chinked with a mixture of one part Portland cement and three parts sand. This is a permanent chinking and will take a stain like the logs if desired. The main room, with fireplace and built-in seat, may be used as the dining room. The windows are casement, which are much cheaper than double-hung windows and can be easily removed and screens inserted for the

VIEW OF LIVING ROOM IN LOG CAMP, NO. 121, SHOWING STONE CHIMNEYPIECE AND BUILT-IN CORNER SEAT.

ventilation of the rooms during the summer.

The main feature is the large open sleeping room upstairs which may be sheltered from wind or rain by duck curtains, and closed up entirely in winter by batten blinds. This sleeping room may be separated into as many small dressing rooms as desired by curtains run on wires or on wooden poles and drawn back when not needed to insure a free circulation of fresh air.

The illustration given here shows the rugged simplicity of the interior, which in spite of its rustic character, holds many possibilities for the making of a comfortable home. The irregular split field stone of the fireplace, with its broad opening and plain

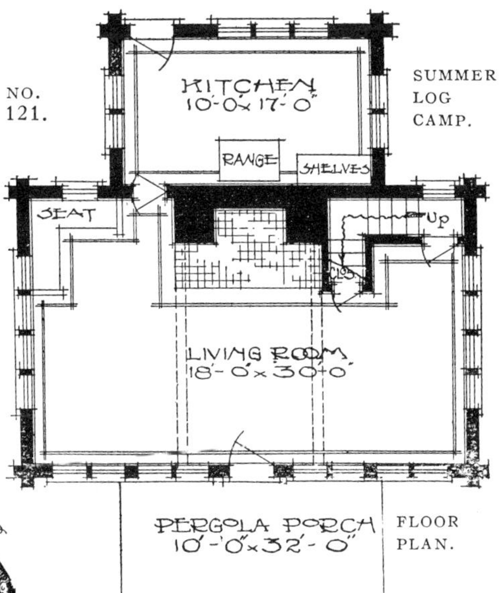

SUGGESTION FOR A RUSTIC GARDEN SEAT.

shelf, the corner seat, the casement windows, with their small, square panes, and the sturdy construction of the logs and boards of walls and ceiling, all combine to form a picturesque background for whatever furnishings may be introduced. Rustic chairs and tables would, of course, harmonize perfectly with this type of construction, but plain oak pieces could be used if preferred, and willow would be both practical and appropriate. On the porch and beneath the trees about the building, rustic seats and benches and small tables would prove a welcome addition to the belongings of the camp, and would extend the boundaries of its comfort

and hospitality. We are showing here sketches of a sheltered seat and arbor, which may serve to suggest other possibilities along the same lines. The putting up of such structures would be a delightful task for the boys of the party, who would no doubt welcome a chance to exert their ingenuity and muscles in such an effective way and prove their skill as builders and carpenters. In fact, a camp of this sort will be found to afford endless opportunities for the development of all those outdoor tasks and pastimes in which labor becomes a source of wholesome joy.

A DECORATIVE GRAPE ARBOR WITH FLOWERS PLANTED FROM POST TO POST

LOG BUNGALOW FOR SUMMER USE, WITH COVERED PORCH AND PARTIALLY OPEN SLEEPING ROOM

Published in The Craftsman, August, 1911.

SUMMER BUNGALOW OF LOGS NO. 122.

VERY like the one previously shown is this camp, both in purpose and general construction. Here, however, the logs are placed horizontally and a long porch is provided, covered, like the main roof, with Ruberoid. The stone fireplace and chimney add to the comfort and picturesqueness of the building, and the upper apartment, being open at both ends, provides an airy place to sleep.

From the porch a door leads into the large living room, and on each side of the door are groups of casement windows with small square panes. Similar windows are also placed in the opposite wall and on each side of the fireplace. In one corner is the staircase leading up to the big, airy room above. A wide opening from the living room leads into the kitchen, which is fitted with long shelves and a stove, and like the living room, has a door opening onto the porch and another door at the rear. A closet is provided in the space beneath the stairs.

The interior view shows one corner of the living room with its stone chimneypiece and massive log construction. The heavy central

PERGOLA PORCH OF SPLIT STONE AND LOGS·

LOG BUNGALOW WITH PARTIALLY OPEN SLEEPING ROOM

PART OF BIG LIVING ROOM IN LOG BUNGALOW NO. 122, SHOWING STONE CHIMNEYPIECE AND CASEMENT WIN-
DOWS, AND EFFECT OF PRIMITIVE COMFORT RESULTING FROM THE WISE USE OF STRUCTURAL FEATURES.

log that runs across the ceiling is in reality two joined in the middle and supported by a post of hewn wood, like the pillars of the porch.

As to the porch itself, pergola construction could be used if preferred, with pillars of split stone, as suggested in the detail sketch, and another porch might be added at the rear of the bungalow if desired. In fact, like all the designs in this book, the plans and the construction may be modified to suit the special requirements of the owner.

The use of vines about the stone chimney and pillars of the porch will add greatly to the charm of the building, and make it even more definitely a part of its surroundings.

In this bungalow, as in the one previously described, box-seats can be built in around the rooms, providing useful storage space for the winter, and rustic chairs and tables, simple and easily made, will prove both an economical and harmonious form of furnishing.

With such a camp, hospitality can be extended indefinitely, for with a living room and kitchen, tent bed-

rooms, hammocks and sleeping bags will afford accommodations for week-end parties.

Since oxygen has proved such an important factor in the prevention and cure of disease, this bungalow, with its open upper story, would be just the thing for a consumptive patient or anyone whose health necessitated the greatest possible amount of fresh air, day and night. Its simplified interior would entail little work, leaving ample time for outdoor life.

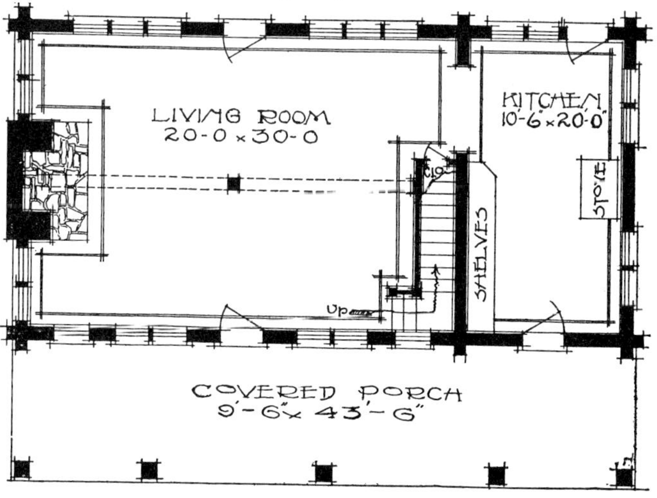

FLOOR PLAN OF LOG BUNGALOW: NO. 122.

105

CRAFTSMAN COUNTRY SCHOOLHOUSE OF LOGS

Published in The Craftsman, July, 1911. CRAFTSMAN LOG SCHOOLHOUSE: NO. 119.

THE rural schoolhouse has in many instances grown into the poorest imitation of city educational institutions, in no way suited to the rural life and environment of farm boys and girls. In most cases as it exists today it not only does not fit them to understand, appreciate and make good in farm life, but actually creates a spirit of discontent with country existence and distaste for real work of any sort. This is a disaster not only to the community, but to the nation, to say nothing of the boys and girls.

America must, for progress' sake, have good country schools, suited to rural conditions. We must have townships that are successful without relation to cities, and people who are contented to dwell in the townships. How to bring this about is one of the most important economic questions of the times. It has seemed to us that something toward this end might be accomplished through the right kind of schools—schools that might become, as did the guildhalls of Mediæval times, the center of a widespread general activity and progress. Why make our schoolhouses such dull, uninviting spots that children must be driven into

them and parents never enter? Why not build schools which will develop the community spirit and definitely prepare the pupils for the kind of lives they are most likely to live? The school should suggest that work, if well done, is not drudgery, but one of the greatest factors in the betterment and uplifting of humanity.

Believing that what our country life needs so vitally is better social, economic and educational advantages, we are showing here two schoolhouses, each designed to be of service to every resident of the district where it is built.

The smaller schoolhouse, No. 119, is made of logs dressed on two sides so that they fit together—the inside and outside left round. The chinking is of cement mortar, which is permanent and takes a stain with the logs, if staining is desired.

The direction of the light is from casement windows at the back and left, and the teacher's desk is placed where full view is had of the two cloak rooms, which are provided with lavatories. Bookcases, closets and blackboards are arranged for in the main room. In rural schools all grades must be accommodated in

CRAFTSMAN COUNTRY SCHOOLHOUSE OF LOGS

one room, so low tables have been set in a bright corner for the little ones. Desk room is provided for forty-two students, and when lectures are given that interest the community at large the kindergarten table can be removed and extra chairs placed around the room, greatly increasing the seating capacity.

Both schoolhouses are planned to be heated and ventilated with a Craftsman fireplace-furnace, which is so simple in management that it can be taken care of by the children themselves. A great advantage of such heating is that fire can easily be kept over night so that the schoolroom will be warm in the morning.

The view of the interior shows the corner of the classroom in which the Craftsman fireplace-furnace is built. The recessed shelf and decorative placing of the irregular stones in the chimneypiece, and the massive effect of the logs make the interior very interesting from a structural standpoint, and the arrangement of the built-in bookcases, with their small glass panes, the casement windows, blackboards and closets, should prove practical and convenient.

Both this schoolhouse and the one on the next page are very simple in design, and could be enlarged to meet the special needs of the community. They suggest, however, what may be accomplished along these lines, and if they help to awaken keener interest in the important subject of rural education and to stimulate a desire for more practical, comfortable and beautiful country schools, they will have served their purpose.

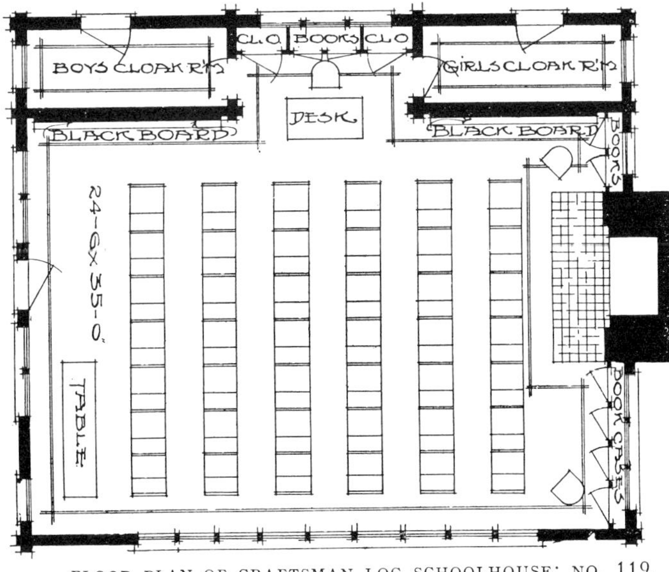

FLOOR PLAN OF CRAFTSMAN LOG SCHOOLHOUSE: NO. 119.

Published in The Craftsman, December, 1908.

VIEW OF THE CLUBHOUSE AT CRAFTSMAN FARMS, SHOWING DECORATIVE EFFECT OF LOG CONSTRUCTION ON THE LOWER STORY AND THE PLASTER PANELS ABOVE. NOTE THE GROUPING OF THE WINDOWS AND THE DECORATED PANELS BETWEEN THE CASEMENTS IN THE DORMER. THE SLOPE OF THE HILL AT THE BACK SHOWS THE POSSIBILITY OF BUILDING THE BASEMENT ROOMS AS DESCRIBED.

A COUNTRY CLUBHOUSE THAT IS BUILT LIKE A LOG CABIN

WE have given the design of the Clubhouse at Craftsman Farms for the use of country clubs that may find such a plan desirable. As we use it ourselves, it will be the general assembly house of the whole colony, so planned that meals may be served either indoors or out on a big veranda, according to the weather, and where meetings, lectures and entertainments of all kinds may be held by the people staying at the Farms and accommodation provided for guests invited from the outside. For our own purpose, no form of building is so suitable and desirable as a low, roomy house built of logs, and we imagine that many a country club will find that similar uses and surroundings seem to demand a building of this character.

As will be seen by comparison of the exterior view of the house with the plan of the lower floor, there are three main divisions in the building, indicated in the perspective drawing by the projecting ends of the logs which form the log partition between the reception room and sitting room

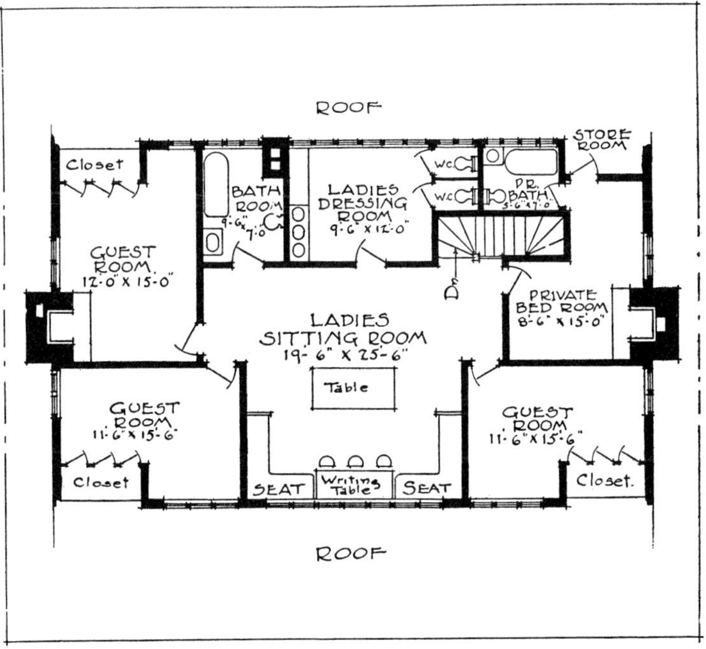

SECOND STORY FLOOR PLAN.

and kitchen on the one side, and serve as the outer wall of the house on the porch side. The width of this porch is the same as the width allowed for the sitting room and kitchen and the center of the building for the whole length is taken up by the reception room, which will be used for the assembly room or the indoor dining room, as seems necessary. The porch will be used as an outdoor living room or dining room, as the case may be, and the little sitting room at the back is meant for guests who may wish some place apart from the general assembly room for a quiet chat with a few friends.

The upper floor is divided into guest rooms, with a comfortable sitting room for ladies

FIRST STORY FLOOR PLAN.

and a dressing room and two bathrooms, so that there is not only accommodation for transient guests but room for a few guests who may wish accommodation over night or for several days at a time.

The smoking room and dressing room for men are placed below the main floor, as in the case of the building at Craftsman Farms the ground slopes sufficiently away from the back of the house to allow ample accommodation for these basement rooms. This slope is sufficiently steep to expose the stone foundation to a depth of seven or eight feet, so that anyone entering the smoking room from the outside comes in on a level instead of going down as into a basement. Flower boxes placed between the pillars around this end of the porch will afford some protection where the slope is most abrupt.

As will be seen, the design of the house is very simple, the effect of comfort and of ample spaces depending entirely upon its proportions. The big sweep of the low pitched, widely overhanging roof is broken by the broad shallow dormers, which not only give sufficient additional height to make the greater part of the upper story habitable, but also adds much to the structural charm of the

building. As the walls of the upper story are of plaster, the logs being used after the manner of half-timber construction, the ends of the dormers are also of plaster and plaster panels divide the groups of casement windows.

These plaster panels form one of the most interesting features of the house because they put into effect our idea of a form of exterior decoration that shall be symbolic of the house itself and the environment in which it stands. Roughly modeled in low relief, are figures symbolizing the life and industries of the farm. Dull colored pigments will be used to emphasize these figures and to add a definite color accent to the house, but the pigments will in all cases come into harmony with the natural tones of wood, stone and earth. These panels form the sole decoration that exists purely for the sake of decoration. For the rest, the beauty of the house depends entirely upon structural features; upon the casement windows, which are all uniform in size and are so arranged as to form long horizontal lines; upon the use of the logs and of stone in the foundation and the chimneys and upon the color harmony of the whole in relation to the prevailing tones of the landscape.

UPSTAIRS SITTING ROOM, SHOWING THE WRITING TABLE AND SEATS IN THE DORMER.

A PLAIN LITTLE CABIN THAT WOULD MAKE A GOOD SUMMER HOME IN THE WOODS

Published in The Craftsman, November, 1908.

VIEW OF THE FRONT AND SIDE, SHOWING CASEMENTS **HIGH** IN THE WALL.

ONE of the features at Craftsman Farms is the housing of guests, students and workers in small bungalows or cabins scattered here and there through the woods and over the hillside, standing either singly or in groups of three and four in small clearings made in the natural woodland. Therefore they are designed especially for such surroundings and are most desirable for those who wish to build inexpensive summer or week-end cottages for holiday and vacation use. Of course, any one of the plans would serve perfectly well for a tiny cottage for two or three people to live in, but the design and general character of the buildings is hardly adapted to the ordinary town lot and would not be so effective in conventional surroundings as in the open country.

The cottages built at Craftsman Farms are meant first of all to live in and next to serve as examples of a variety of practical plans for small moderately priced dwellings designed on the general order of the bungalow. They will be built of stone, brick, or any one of a number of our native woods suitable for such construction and will be as comfortable, beautiful and interesting as we can make them, each one being specially planned for its own use.

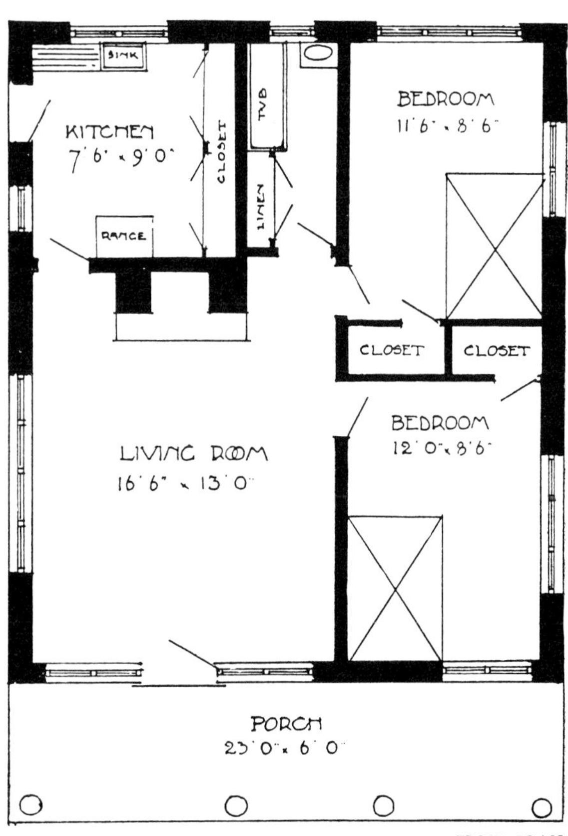

FLOOR PLAN.

111

Published in *The Craftsman, July, 1904.*

A BUNGALOW BUILT AROUND A COURTYARD FACING THE WATER: VIEW OF THE FRONT, SHOWING ENTRANCE, TERRACE AND CHIMNEYS, AND THE RELATION OF THE HOUSE TO ITS SITE.

A BUNGALOW BUILT AROUND A COURTYARD FACING THE WATER

ONE of our earliest designs is shown in this bungalow, which has proven very popular for summer homes, especially where they are built on the shore of a lake or river; for the chief characteristic of the design is an inner court, or *patio*, which looks directly out upon the water. The bungalow is built around three sides of this courtyard,—an arrangement which carries with it a suggestion of the old Mission architecture of California.

The original design was for a house with shingled walls, but the construction is equally suitable for stone, brick, or concrete. The material chosen, of course, would depend entirely upon the locality and the taste of the owner. Were we designing it now, we would probably suggest concrete, as the form of the house, with its straight walls and simple lines, is well suited to this material, and also because this method of construction is comparatively inexpensive as well as substantial and durable. If the walls were finished with rough plaster or pebbledash surface, the effect would be admirable, especially for the woods, if a little dull green pigment were brushed on irregularly, giving a general tone of green that yet is not a solid smooth color.

The central court as shown here is paved with stone, but this would be only in case of stone or shingle construction. For either brick or concrete it would be best to pave the court with cement colored a dull red and marked off into squares. This has much the appearance of Welsh quarry tiles and is much less expensive. Provision has been made in the center of the court for a basin, in the middle of which a pile of rocks affords opportunity for a fountain or trickling cascade, while the pool furnishes an admirable place for the growth of aquatic plants. The court can either be paved clear up to the pool as shown in the picture, or the pavement may stop just outside the pillars, leaving the center of the courtyard for turf. In either case the *patio* is meant to be furnished for use as an outdoor living room, such as is so frequently seen in the courtyards

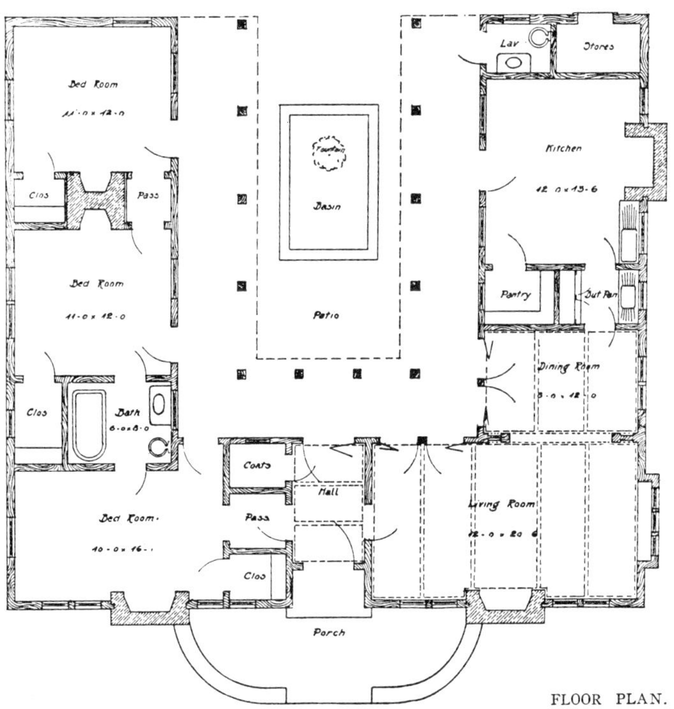

FLOOR PLAN.

of California houses. If the house is built for a camp in the woods, the pillars around this courtyard would best be made of peeled logs left in the natural shape and stained back to the color of the bark. For more conventional use, heavy round pillars of concrete or of wood painted white would naturally be used. These details, however, are always ruled by the locality, the materials used for building and the taste of the builder.

The arrangement of the interior is very simple, as from the entrance hall one turns toward the right into the living room, which occupies half the front of the building. Just back of the living room in the wing is the dining room and back of this again is the kitchen. Turning to the left from the hall, a small passage leads to one of the bedrooms, and the other two bedrooms and the bathroom occupy the whole length of the wing. All of these rooms open out upon a central court and all are lighted from the outside by casements set high in the wall. Fireplaces are plentiful, the chimneys being so arranged that one is allowed for each bedroom and one for the living room. This being almost opposite the dining room, or rather alcove, serves for that room as well.

113

VIEW OF THE COURTYARD LOOKING OUT UPON THE WATER. THIS MIGHT BE USED AS AN OUTDOOR LIVING ROOM AND FURNISHED WITH TEA TABLE, CHAIRS, HAMMOCKS AND RUGS. BEING SO COMPLETELY SHELTERED FROM THE WIND IT SHOULD MAKE A VERY PLEASANT AFTERNOON LOUNGING PLACE IN WARM WEATHER.

A RUSTIC CABIN THAT IS MEANT FOR A WEEK-END COTTAGE OR A VACATION HOME

Published in The Craftsman, November, 1908.

FRONT VIEW OF CABIN, SHOWING DECORATIVE USE OF TRUSS IN THE GABLE.

THIS is another example of the cottages built at Craftsman Farms and is somewhat larger than the stone cabin shown on page 81, as it contains a bathroom and a recessed porch which serves as an open air dining room, in addition to the living room, two bedrooms and kitchen provided in the smaller cottage.

The walls are sheathed with boards eight or ten inches wide and seven-eighths of an inch thick. A truss of hewn timber in each gable, projecting a foot and a half from the face of the wall, not only gives added support to the roof, but forms a decorative feature that relieves the extreme simplicity of the construction. The casement windows are all hung so they will swing outward and are mostly small and set rather high in the wall. At the ends of the building these casements are protected by simple shutters, each one made of two wide boards with either heart shaped or circular piercing. These solid shutters provide ample shelter in severe weather.

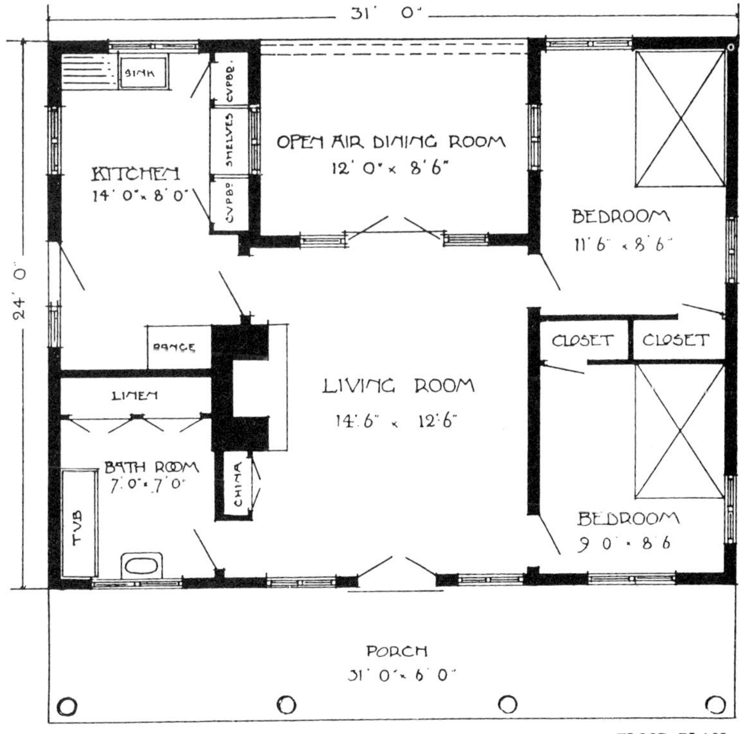

FLOOR PLAN.

115

A BUNGALOW DESIGNED FOR A MOUNTAIN CAMP OR SUMMER HOME

Published in The Craftsman, March, 1905.

REAR VIEW OF BUNGALOW, WITH VERANDA LOOKING TOWARD THE WATER.

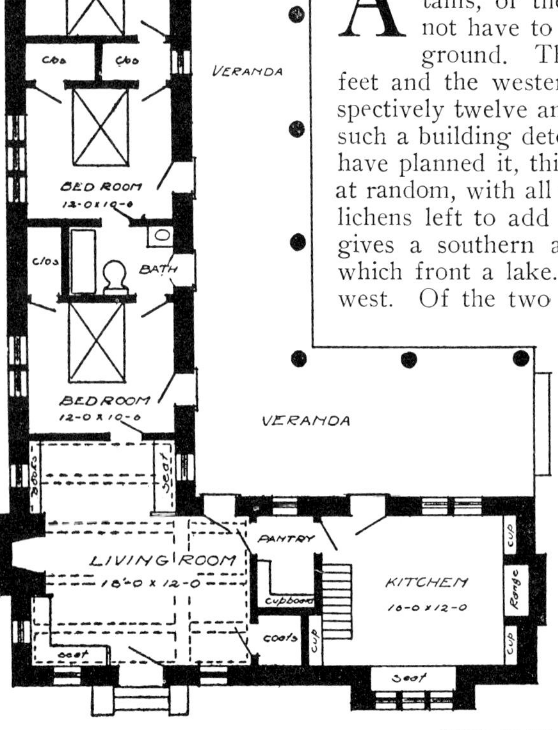

FLOOR PLAN.

A S this bungalow is meant either for the woods, the mountains, or the open country, where the cost of land does not have to be considered, it spreads over a good deal of ground. The eastern wing has a frontage of sixty-four feet and the western of forty-four feet, the verandas being respectively twelve and ten feet. Also the probable environment of such a building determines the character of the exterior. As we have planned it, this bungalow is built of rugged field stones set at random, with all the weather stains and accretions of moss and lichens left to add to the color value. The site suggested here gives a southern and western exposure to the wide verandas which front a lake. The building itself faces toward the northwest. Of the two wings, the eastern, containing the bedrooms, extends into the wooded portion of the land in order to insure protection and coolness; while the west wing looks toward the lake.

The interior of this bungalow is divided into a living room, a kitchen and three bedrooms. The living room is large and comfortably arranged, the idea being to give it a character in harmony with the plan, purpose and exterior effect of the building.

The kitchen is planned so that meals may be served in it in bad weather. Ordinarily the meals would be served in the sheltered corner of the veranda. The whole eastern wing is given up to the bedrooms which are all entered from the veranda, and overhead is a large storage attic.

116

A MOUNTAIN CAMP OR SUMMER HOME

FRONT VIEW OF BUNGALOW, SEEN FROM THE LAND.

CHIMNEYPIECE AND FIRESIDE NOOK IN THE LIVING ROOM. NOTE THE USE OF LOGS FOR OVERHEAD BEAMS AND OF WIDE V-JOINTED BOARDS FOR THE WALLS AND SEAT.

A CONVENIENT BUNGALOW WITH SEPARATE KITCHEN AND OPEN AIR DINING ROOM

FRONT AND REAR VIEWS OF COTTAGE, THE FIRST SHOWING RECESSED ENTRANCE PORCH AND THE SECOND THE OPEN-AIR DINING ROOM WHICH SEPARATES THE KITCHEN FROM THE MAIN PART OF THE HOUSE.

A BUNGALOW WITH OPEN-AIR DINING ROOM

FOR any place, whether mountain or valley, that is really " in the country," the best form of summer home is the bungalow. It is a house reduced to its simplest form, where life may be carried on with the greatest amount of freedom and comfort and the least amount of effort. It never fails to harmonize with its surroundings, because its low broad proportions and absolute lack of ornamentation give it a character so natural and unaffected that it seems to sink into and blend with any landscape. It may be built of any local material and with the aid of such help as local workmen can afford, so it is never expensive unless elaborated out of all kinship with its real character of a primitive dwelling. It is beautiful, because it is planned and built to meet simple needs in the simplest and most direct way; and it is individual for the same reason, as no two families have tastes and needs alike.

The bungalow illustrated here is designed on the purest Craftsman lines. The material we have suggested is cedar shingles throughout with a foundation and chimney of rough gray stone. No cellar is provided, but the walls have a footing below the frost line and space under the floor for ventilation. The building is in the form of a T, the main portion covering a space twenty-four by forty feet and the extension at the back fourteen by thirty-six feet. The low-pitched, widely overhanging roof gives a settled, sheltered look to the building, and this is emphasized even more by the deeply recessed porch in front, which is meant to be used by a small outdoor sitting room. The porch be-

tween the kitchen and the main part of the house is really a portion of the extension left with open sides and is intended for an outdoor dining room that shall be sufficiently sheltered

FLOOR PLAN.

A BUNGALOW WITH OPEN-AIR DINING ROOM

END OF LIVING ROOM, SHOWING BALCONY, FIRESIDE NOOK WITH CHIMNEYPIECE AND ARRANGEMENT OF STAIRCASE.

ARRANGEMENT OF CUPBOARDS, WORK SHELF AND WINDOWS IN KITCHEN.

from storms to allow the outdoor life to go on through any sort of weather.

The living room occupies the whole center of the house, except for the recessed porch in front, and it is one of the best examples of the Craftsman idea of the decorative value that lies in revealing the actual construction of the building. Everyone knows the sense of space and freedom given by a ceiling that follows the line of the roof. It seems to add materially to the size of the room and when it is of wood it gives the keynote for a most friendly and restful color scheme. In this case the whole room is of wood, save for the rough gray plaster of the walls and the stone of the fireplace. A balcony runs across one side, serving the double purpose of recessing the fireplace

A BUNGALOW WITH OPEN-AIR DINING ROOM

into a comfortable and inviting nook, and of affording a small retreat which may be used as a study or lounging place, or as an extra sleeping place in case of an overflow of guests, or even as a storage place for trunks. Its uses are many, but its value as an addition to the beauty of the room is always the same.

The sleeping rooms, four in number, occupy the two ends of the main building. They are all of ample size for camp life, and are plastered, walls and ceiling. The dining porch is one of the most distinctive features of the bungalow. It occupies just half of the extension and completely separates the kitchen from the main part of the house. The kitchen is well open to air and light. Instead of a pantry the whole of one side is occupied by cupboards amply supplied with shelves and drawers.

121

CRAFTSMAN GARDENS FOR CRAFTSMAN HOMES

A CRAFTSMAN house should be surrounded by grounds that embody the Craftsman principles of utility, economy of effort and beauty. All these qualities it is possible for the average man to achieve in his garden by a little careful study and skilful planning. The majority of home owners to-day are people who must necessarily depend upon their own efforts for taking care of and beautifying their home grounds. As far as the men are concerned, they are as a rule workers in the city who could afford to give perhaps a part of Saturday and all day Sunday to any garden they had, with an occasional hour in the morning and the evening and holidays thrown in. This, of course, means that their gardens must be planned in such a way as to require the minimum amount of care and stand the maximum amount of neglect. In answer to the obvious question: "Since the time I could spend upon it is likely to be limited, could I really have much of a garden?" the answer is emphatically, "You can if you wish. You can have a most considerable garden of vegetables, flowers, fruit and berries that will quite fulfil the purposes of beauty and utility and give you a splendid outlet for your natural desire to grow things." The amount of ground you have is a ruling factor, of course, in your plans, but even on the smallest suburban lot, say sixty by one hundred feet, perhaps less, a very satisfactory garden scheme can be worked out.

In order to illustrate practically just what can be done, we have taken four of our most popular designs for Craftsman houses and have made garden plans for them in which the most economical use of the surrounding land has been taken into consideration, and in which we have had regard also for beauty. In house number one we have taken a plot approximately seventy-five by one hundred feet and put on it a house that is about forty feet square, and we have pictured it as it would appear in the early spring. As will be noted, we have provided for a vegetable garden, a drying space, an orchard, a good-sized lawn and flower borders. In laying out the part devoted to vegetables we have suggested a large number of paths. These paths are almost a necessity. While they cut down the space, they make it possible for the home owner to hoe his vegetables without going up to his ankles in mud, and thus the garden is likely to get much more attention. The space as given does not seem large. It will, however, provide more vegetables than the average person would imagine, and would certainly grow sufficient of the staple vegetables to keep a family of four or five well supplied throughout the summer.

In choosing the vegetables you will plant, and, in fact, in considering the entire garden scheme, it is best to be careful not to plan for more than you can really take care of. Agree with yourself that you will be faithful to your garden; decide just how many hours you are sure you will be able to spend each week upon it, and err on the small side in making your estimate, rather than on the larger. Do not put into your vegetable garden things which will require a large amount of cultivating throughout the season, such as celery, which has to be banked up. Choose the standard things such as peas, beets, beans, green onions, carrots, spinach, radishes, limas, parsley, turnips, that practically can be had for the trouble of sowing, harvesting and a small amount of labor each week. Tomatoes, lettuce and asparagus require a little attention and should be added only after considerable thought. It is better not to have them than to have them come to nothing through neglect. You can have corn and squash and cabbage,

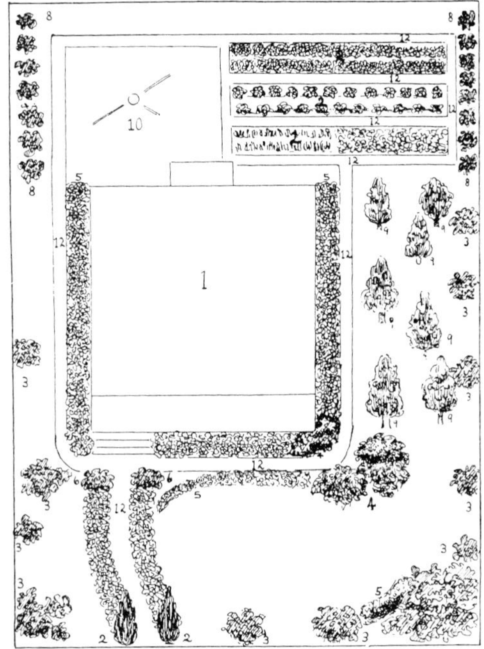

PLAN FOR PLANTING SPRING GARDEN: NO. 1.

CRAFTSMAN HOUSE WITH SPRINGTIME GARDEN IN BLOOM: YELLOW PREDOMINATING: NO. 1.

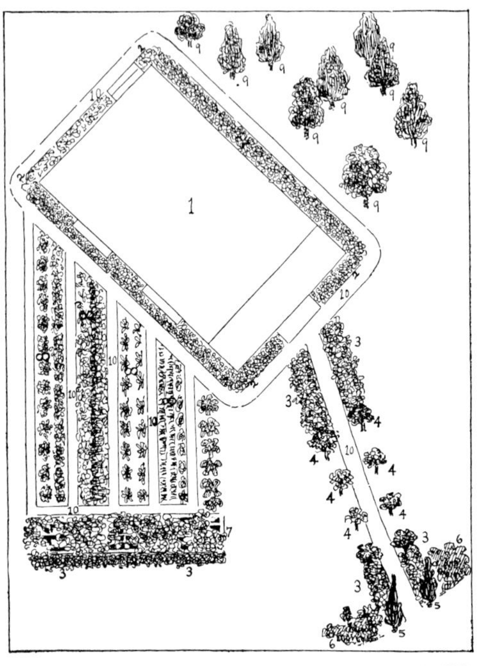

PLAN FOR PLANTING EARLY SUMMER GARDEN: NO. 2.

small fruits for the Craftsman garden are gooseberries and currants. These bushes ask practically no attention. Raspberries are possible, but they require cutting down each year. Blackberries should be avoided because they have a tendency to run wild. The plan provides, as can be seen, a good piece of lawn close to the house. It is best that this should be kept practically open and free from small flower beds or shrubs, as these are troublesome when the lawn is being mowed.

The character and color of the house itself must be very carefully taken into consideration in choosing and planting the flowers, shrubs and trees. It goes without saying that a house should have some trees about it; if there are any already on the ground, so much the better; if not, of course, they must be furnished. Trees of a very satisfactory size, quite large enough to be really impressive, can be bought from any nursery, and if the home owner can afford nothing else he should at the outset afford several good trees. Evergreens, such as cedars, spruces, firs and arborvitæ are most satisfactory because they decorate the grounds the year through. On the plan we have indicated a cedar tree at each side of the front gate. These grow quite tall and have a pyramidal shape that suits them especially for flanking the gateway. Maples grow rather quickly and one placed close to the house might be added to this plan, to take away any sense of bareness from the façade. Birches, because of their beautiful white bark, are decorative even in winter, and one ought to be included among the trees planted. Dwarf Japanese cut-leaf maples have a beautiful red foliage in spring and fall, and a place should be found for at least one where it will be seen against evergreens, if possible.

The flower garden should be planned with a view to its harmonizing in color with the house. The first illustrated is brown with a dark red roof. Success in making the colors of the flowers harmonize with the house is merely a matter of careful thought and planning. One can have from flowers almost as many colors as a painter can mix on his palette, and one can have them from early spring until late fall, and in the winter one can have shrubs with beautiful red, yellow or green branches. What are known as hardy herbaceous plants are the most popular ones now, and justly so. They are the best ones for a Craftsman garden because they mean the smallest amount of

and perhaps muskmelons and watermelons, too, if your space permits. If you wish to add potatoes you must be able to provide considerable land and time for them.

It is well to bear in mind that horticulture specialists are all the time studying to produce varieties of vegetables, fruits and flowers that will stand bad conditions and neglect and be free from pests. It is wise to get the catalogues of good seedsmen, read them carefully for suggestions, because they are usually reliable, and select those varieties of flowers and vegetables which are quoted as most hardy.

Fruit and some small berries can be included in the garden of a Craftsman house. Recently very satisfactory dwarf fruit trees have been developed. You can get apples, peaches, cherries, plums and nectarines. They grow to about six feet and are very compact of form, and produce for their size a large amount of perfect, good-flavored fruit. They are especially suited to a Craftsman garden because, though like all fruit trees they must be sprayed and pruned, these processes involve the smallest amount of labor and can be done from the ground instead of from ladders. In plan number one we have placed these trees on the south side of the house where they will get the largest amount of sun. The best

AN EARLY SUMMER CRAFTSMAN GARDEN IN ROSE, LAVENDER, BLUE AND WHITE: NO. 2.

trouble, and because they are likely to survive the largest amount of neglect. After they are once put out they stay in their places forever, and all they need is to be raked around with the hoe occasionally and to have their roots thinned out when they have begun to grow too thickly. Even when not in bloom they furnish decorative foliage to cover the bare earth. In planting the flowers make a careful selection so that you may have a succession of bloom and so that the colors of the flowers shall not clash with each other or clash with the house. Be careful not to put the magenta flowers against pink ones for example, or to have on the porch climbing roses that will not harmonize with the red of the roof, or purple against pink.

The suggestions we give for planting the garden of house number one will bring a general impression of a cheerful yellow all over the garden. In the garden scheme flowering shrubs must be included, and we have suggested here forsythia or golden bell as the most important shrub. To assist in giving the yellow effect we have included daffodils sprinkled thickly in the borders, also red and yellow tulips, yellow iris and yellow crocus. At each side of the steps we have placed a yellow peony. With these flowers in predom-

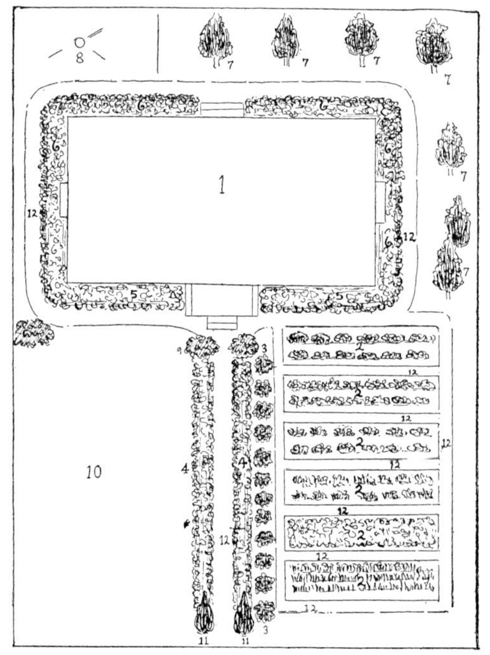

PLAN FOR PLANTING LATE SUMMER GARDEN: NO. 3.

inance an especially bright and sunny effect will be produced in the springtime. In the beds there will, of course, be other hardy annuals showing their foliage to fill in the bare spots. These will come out later, but they naturally also should be planted with an understanding of the combination of color they will make at their period of bloom, and its relation to the house. The plot surrounding this house is seventy-five by one hundred feet, room enough for a small garden.

House number two is built of grayish brown stone and brown shingles, and has a dull-green roof. The plot on which it is located is of slightly irregular shape, as plots usually are in the better class of properties. It is about one hundred and fifty by two hundred feet and slopes slightly up from the northeast to a level space, on the edge of which the house is placed, facing southeast in order to get the sunlight in the living room, dining room and main bedrooms.

In making the plan for the flower planting we have had in mind the general appearance of the place in early summer. These are the months when one can expect to get the best out of one's roses. A delightful rose for a Craftsman garden is the Japanese variety usually called rugosa, which seems to be proof against all floral ailments. It produces flowers somewhat like the wild rose, only larger and richer in color, and has a thick, somewhat lustrous foliage that makes it very satisfactory as a shrub as well as a flower. It is being constantly developed, and the newer proved varieties are sure to be satisfactory to the Craftsman gardener. It produces large red seed-pods that are extremely decorative in the fall. Rugosa roses can be planted freely among the shrubs. A climbing rose is always a cheerful decoration to a house. It softens the lines and gives shade if allowed to run over the porch. Some varieties of climbing roses bloom with an almost miraculous profusion. As the roof of the house is a dark green, we would suggest in this case a deep pink climbing rose. Standard roses are those that have been grafted to the top of a sturdy trunk, and usually stand two or three feet high, bushing out at the top. These can be planted at the edges of a walk, as we suggest in this plan. They have a note of formality that is not too strong to harmonize with a Craftsman house.

The choice of shrubs offered for this time

A LATE SUMMER CRAFTSMAN GARDEN, WITH FLOWERS AND VEGETABLES SIDE BY SIDE: NO. 3.

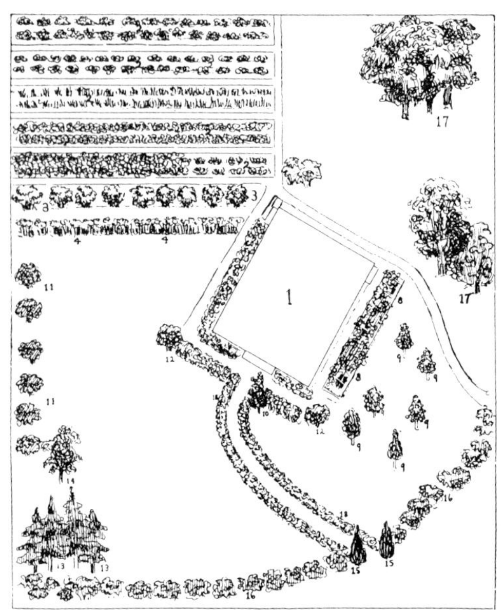

PLAN FOR PLANTING FALL GARDEN: NO. 4.

of the year is quite generous. We have in mind a scheme of coloring largely lavender and white. The key to this scheme will be set in shrubs by the lilacs which can be had in both white and lavender. There are Japanese snowballs, mock orange and spiræa for the note of white in the shrubs. Among the flowers, columbine, iris, forget-me-not and lily-of-the-valley will contribute to the general lavender and white effect, and will come in bloom in this period.

Grapes a Craftsman gardener can have without much trouble, and a grape arbor is included in the second plan. The vegetable garden is placed to the side and is screened from the road by the grape arbor, and gooseberry and currant bushes have been planted along another side to hide it partially from the main pathway to the house. If borders are placed on the lawn side of the arbor and the row of bushes the effect will be very satisfactory. There should also be borders in front of the house, and we have suggested that they run part of the way down the path from the house on both sides, and part of the way up from the gate on both sides, the standard roses serving to join the two effects of bloom. Poplar trees are of interesting shape, making slim pyramids, and are suitable to plant, as we have indicated in this plan, at the sides of the gateway. A low hedge of privet might be placed at the edge of the lawn to separate it from the public roadway or sidewalk. Dwarf fruit trees would be effective on the slope at the east side of the house, and evergreens might be clustered behind the beds at the gateway and in the front corners of the plot. A maple tree and perhaps a birch might be planted close to the house.

An interesting arrangement of flowers for midsummer, that would harmonize with this house would be one emphasizing the blues and whites. This would make the garden seem cool during the hot July and August days when one prefers to have the red and yellow out of sight. The larkspurs, the campanulas or bell flowers, the aconitums or monkshoods, and the platycodons will make a good show of blue and white at this time. And among the shrubs deutzia and blue spiræa are in blossom.

Plan number three is made for the effect of late summer. One of the important features of this plan is the placing of the vegetable garden in front of the house. A properly kept vegetable garden is in its way as beautiful as a flower garden, and by treating it decoratively and letting it have here and there a few clumps of flowers, it can be made a very charming spot indeed. It will be in conformity with the Craftsman spirit that so essential a part of the home as the vegetable garden need not be hidden. In late summer this garden will have its vegetables well toward maturity, and if corn has been planted it will be showing its decorative foliage. As one method of marking the boundary of this vegetable garden we have placed gooseberry and currant bushes between it and the pathway to the house. The floral scheme consists of long borders at the edge of the path from the gate, and borders around the house. A few annuals, such as nasturtiums, poppies, asters and cosmos require so little attention that they can be used profusely in a Craftsman garden. This late summer plan calls for poppies at the front of the house. In the long borders beside the main walk the predominant flower is golden glow. Hollyhocks have been indicated in the beds at the side of the house. Unless somewhat protected from heavy winds, these are likely to be damaged, and so a sheltered location such as the one indicated is best. The plot for this garden is seventy-five by one hundred feet.

The substantial house we have chosen in

A BRILLIANT FALL CRAFTSMAN GARDEN, FOR LITTLE LABOR AND MONEY: TONE RED AND PURPLE: NO. 4.

this case is of cement with brownish yellow stone and a brown roof. The general effect of the flowers in the garden will harmonize well with this house, since the predominant colors are a warm yellow and white. Rose of Sharon is a good-sized shrub that blossoms in late summer. The white variety would be best for this planting. The house faces the north, and the dwarf fruit trees are placed to the south and west. The evergreens and other trees could be planted at the edges of the plot with the shrubs in between. An interesting arrangement would be to have little round box bushes flanking the gateway, and Rose of Sharon on each side of the pathway in front of the house.

The fourth garden plan contains suggestions for planting with fall effects in mind. Here again the house is on an irregularly shaped piece of property and on the brow of a slight slope. This piece of ground is about two hundred and twenty-five feet by one hundred and seventy-five feet, and the house is placed to face the northwest. The garden occupies ground to the northeast and is partly screened by gooseberry and currant bushes, before which is a flower border. In the front of the house are two borders with a path between, the one closest the house being filled with nasturtiums that keep up their bloom until frost, and the other devoted largely to red, white and yellow chrysanthemums. The path down to the gate is also fringed with nasturtiums.

One of the most beautiful fall flowers—an annual, by the way—is salvia. It is so wonderful in color that one can hardly afford to do without it, but it must be started indoors in March in "flats." Its color is so decided that it kills nearly everything else, and so should be very carefully handled. We have indicated salvia on one side of the house, where it will be seen almost alone and not clash with other flowers. Cosmos will last till frost and might be planted in the border shielding the vegetable garden, in some strong deep reds that would stand the proximity of the brilliant salvia. The grape arbor is placed to the southwest of the house, and the orchard of dwarf fruit trees on the slope to the southwest. What is known as Japanese barberry, that turns an exquisite deep red in the fall, makes a hedge of moderate height. This might be used to divide the lawn from the roadway. About the only shrub that can be

counted on at this time of the year is the hardy hydrangea, and we have suggested one placed at each end of the second border before the house. The Japanese cut-leaf maples have a gorgeous foliage in fall, and a good specimen tree of this would be effective near the house.

This is the time of the year when evergreens will be most useful, and a house to be occupied in the fall should have clumps of such trees planted about.

SCHEDULES OF PLANTING FOR EARLY SPRING GARDEN. 1.—House. 2.—Cedars. 3.—Shrubs: forsythia, spiræa, deutzia, etc. 4.—Banks of rose-bushes: Red, yellow and white rugosas. 5.—Border beds of hardy flowers: Daffodils, iris, crocus, tulips in bloom, backed by other flowers such as larkspur, columbine, phlox, bell flower. 6.—Peony bushes. 7.—Plots with vegetables. 8.—Gooseberry and currant bushes. 9.—Dwarf fruit trees in bloom. 10.—Drying ground. 11.—Lawn. 12.—Walks. (See page 168.)

SCHEDULE OF PLANTING FOR EARLY SUMMER GARDEN. 1.—House. 2.—Border: Lily-of-the-valley close to house, phlox. 3.—Border: Sweet-william, iris, phlox, etc. 4.—Standard rose-bushes. 5.—Poplars. 6.—Shrubs: Japanese snow-ball, mock-orange, spiræa, backed by small evergreens. 7.—Grape arbor. 8.—Vegetable garden. 9.—Dwarf fruit trees. 10.—Walks. 11.—Currant and gooseberry bushes. (See page 170.)

SCHEDULE OF PLANTING FOR LATE SUMMER GARDEN. 1.—House. 2.—Vegetable garden. 3.—Currant and gooseberry bushes. 4.—Borders with phlox and golden glow as main flower. 5.—Border with poppies. 6.—Borders with hollyhocks. 6a.—Lilies in the vegetable garden. 7.—Dwarf fruit trees. 8.—Drying yard. 9.—Rose of Sharon. 10.—Lawn. 11.—Pyramidal box bushes at gateway. 12.—Walks. (See page 172.)

SCHEDULE OF PLANTING FOR FALL GARDEN. 1.—House. 2.—Vegetable plot,—corn prominent. 3.—Currant and gooseberry bushes. 4.—Border with red and white cosmos. 5.—Border of salvia. 6.—Border of nasturtiums. 7.—Border with chrysanthemums: Red, white and yellow. 8. — Grape arbor. 9. — Dwarf fruit trees. 10. — Japanese cut-leaf maple. 11. — Shrubs. 12.—Hardy hydrangeas. 13.—Evergreens. 14.—Birch or pin oak. 15.—Pyramidal privets. 16.—Japanese barberry hedge. 17.—Old trees. (See page 174.)

PORCHES, PERGOLAS AND TERRACES: THE CHARM OF LIVING OUT OF DOORS

IN these days when the question of light and air is of so much importance in the planning of the home, the tendency is more and more toward the provision of ample room for as much open-air life as possible. In all the Craftsman houses, as well as in the best modern dwellings of other styles, the veranda, whether open in summer or enclosed for a sun room in winter, is one of the prominent features. Partly for convenience in enclosing with glass if desired, but mainly to insure the pleasant sense of privacy that means such a large part of the comfort of home, these porches or verandas are usually recessed so that they are partially protected by the walls of the house and are further sheltered by the copings and flower boxes. In a front porch which must serve for a sitting room as well as an entrance, the coping, surmounted by flower boxes, acts as a screen and, with the aid of a generous growth of vines, serves as a very satisfactory shelter from the street. Where there is also a garden veranda it can be made into a charming outdoor living or dining room both for summer and for mild days in winter by being so recessed and protected that it is like a summer house or an outdoor room always open to the sun and air.

Outdoor living and dining rooms, to be homelike and comfortable, should be equipped with all that is necessary for daily use so as to avoid the carrying back and forth of tables, chairs and the like, as when the veranda is used only occasionally. It goes without saying that the furniture should be plain and substantial, fitted for the rugged outdoor life and able to stand the weather. Indian rugs or Navajo blankets lend a touch of comfort and cheer, and the simple designs and primitive colors harmonize as well with trees and vines and the open sky as they do with their native wigwams. Willow chairs and settles seem to belong naturally to life in the garden, and with a few light tables, a book rack or two and plenty of hammocks, the veranda has all the sense of peace and permanency that should belong to a living room, whether indoors or out, that is habitually used by the family.

Published in The Craftsman, June, 1905.

COURTYARD AND PERGOLA, SHOWING DECORATIVE EFFECT OF THE CENTRAL SQUARE OF TURF WITH ITS FOUNTAIN, SHRUBS AND ROCKS AND THE COMFORT OF THE VINE-SHADED PORCH WHEN FURNISHED FOR USE AS AN OUTDOOR LIVING ROOM.

Published in The Craftsman, November, 1906.

PORCH THAT NOT ONLY SERVES AS A DESIRABLE ENTRANCE BUT GREATLY INCREASES THE STRUCTURAL INTEREST OF THE FRONT OF THE HOUSE. THE WINDOWS ON EITHER SIDE PROJECT SLIGHTLY FROM THE WALL IN A SHALLOW BAY AND THE ENTRANCE DOOR WITH ITS CASEMENTS ON EITHER SIDE PROJECTS STILL FARTHER. THE PORCH IS COMPARATIVELY NARROW AND THE ROOF IS SUPPORTED BY TWO HEAVY PILLARS OF WOOD PAINTED WHITE, WHICH SERVE TO GIVE ACCENT TO THE DARKER TONES OF THE SHINGLES, EXTERIOR WOODWORK AND STONE FOUNDATION. THE WALLS ARE SHEATHED WITH CYPRESS SHINGLES THAT ARE OILED AND LEFT TO WEATHER, AND THE WOOD-WORK OF THE ROOF, DOOR AND WINDOW FRAMINGS AND BALUSTRADE IS IN A DARKER TONE OF BROWN. THE USE OF SPINDLES FOR EXTERIOR WOODWORK IS SHOWN IN THE BALUSTRADE.

Published in The Craftsman, September, 1906.

ENTRANCE PORCH FITTED UP FOR AN OUTDOOR LIVING ROOM. FLOOR OF WELSH QUARRIES COVERED WITH A LARGE RUG INTENDED TO STAND ROUGH USAGE AND EXPOSURE TO THE WEATHER. THE BEAMED CEILING IS FORMED BY THE EXPOSED RAFTERS AND THE PORCH IS PARTIALLY SHELTERED BY THE PARAPET.

PORCHES, PERGOLAS AND TERRACES

ENTRANCE PORCH TO A HOUSE BUILT OF ROUGH CAST CEMENT. THE WOODEN PILLARS ARE PAINTED PURE WHITE AND ARE VERY THICK AND MASSIVE IN PROPORTION TO THEIR HEIGHT. THE RAFTERS ARE LEFT IN VIEW WHERE THEY SUPPORT THE ROOF AND A HEAVY BEAM RUNNING THE LENGTH OF THE PORCH SERVES TO UPHOLD THE RAFTERS. SQUARE MASSIVE CROSS-BEAMS EXTEND FROM THE PILLARS TO THE WALL, WHERE THE ENDS ARE SUNK IN THE FRAMING OF THE HOUSE. THE FLOOR AND STEPS ARE OF CEMENT COLORED A DARK RED AND MARKED OFF IN BLOCKS LIKE TILES.

Published in The Craftsman, November, 1904.

RECESSED PORCH AT REAR OF HOUSE, SHOWING RELATION OF THE PORTION THAT IS SHELTERED TO THE OPEN TERRACE THAT EXTENDS BEYOND THE ROOF AND GIVES SUFFICIENT SPACE FOR AN OUTDOOR LIVING ROOM THAT IS PARTLY OPEN TO THE SKY.

Published in The Craftsman, August, 1905.

RECESSED ENTRANCE PORCH FURNISHED AS AN OUTDOOR LIVING ROOM THAT IS PARTLY SHELTERED BY THE WALLS AND PARTLY SCREENED BY VINES.

PERGOLAS IN AMERICAN GARDENS

A DOUBLE PERGOLA, VINE-COVERED AND ROSE-GROWN: THE OVERHEAD POLES OF THE PERGOLA ARE OF CEDAR, AND THEIR RUSTIC EFFECT IS IN KEEPING WITH THE PLANTING SCHEME AND IN PLEASING CONTRAST TO THE FORMAL LINES OF THE HALF-TIMBER OF THE HOUSE FROM WHOSE PORCH THE PERGOLA STRETCHES FORTH.

WHATEVER connects a house with out of doors, whether vines or flowers, piazza or pergola, it is to be welcomed in the scheme of modern home-making. We need outdoor life in this country; we need it inherently, because it is the normal thing for all people, and we need it specifically as a nation, because we are an overwrought people, too eager about everything except peace and contentment. I wonder if anyone reading this article has ever in life received the following invitation, "Will you come and sit in my garden with me this afternoon?" I doubt it very much, at least in America. In England this would happen, or in Italy, and I think in Bavaria the people rest in their gardens at the close of the day and grow strong and peaceful with the odor of flowers about them, and the songs of birds. In a garden the silence teaches the restless spirit peace, and Nature broods over man and heals the wounds of the busy world. In essence a garden is a companion, a physician, a philosopher. It is equally the place for the happy, the sorrowing, for the successful, for the despondent.

And so here in America of all things we need gardens, and we must so plan our gardens that we shall live in them, and we must have in them our favorite flowers, long pathways of them, which lead us from gate to doorstep, and we must enter our gateway under fragrant bowers. We must build up arbors for our fruit, rustic shelter for our children, and above all these things our garden, which should be our outdoor home, must surely have a pergola, a living place outdoors that is beautiful in construction, that is draped in vines, that gives us green walls to live within, that has a ceiling of tangled leaves and flowers blowing in the wind, a glimpse of blue sky through open spaces and sunshine pouring over us when the leaves move.

PERGOLAS IN AMERICAN GARDENS

With a pergola in the garden you can no more escape living out of doors than you can avoid swimming in the sea if you happily chance to be living on the edge of the ocean. A pergola focuses your garden life. It is like a fireplace in a living room; it is the spirit of the outdoor environment held in one place to welcome you. It is essentially a place in which to rest, or to play or to do quiet domestic tasks; it is the outdoor home for children, for old folks, a spot in which to dream waking dreams or to sleep happily, or, best of all, for

on the other hand, as in one of the illustrations, it gracefully hides a group of unbeautiful farm buildings. It may lead to a beautiful garden or out to a wonderful view, or it may be the culmination of the garden scheme and furnish the only vista of which limited grounds are capable. It epitomizes modern outdoor life, and its beauty is through simplicity of construction and intimacy with Nature. A pergola inevitably means good simple lines of construction, beautified with vines, hidden with fruit or flowers, and with sunlight in splashes

SHOWING THE USE OF PERGOLAS TO HIDE IN A PICTURESQUE FASHION THE OUTBUILDINGS OF A FARM: THIS PICTURE IS A REPRODUCTION OF THE PERGOLAS AT THE STETSON FARMS, STERLINGTON, N. Y.

romance. For a pergola is a wonderfully inspiring spot in twilight, or when moonlit.

The outdoor living place is suited equally to any landscape or climate. It can be adjusted to any kind of architecture. It can be built directly with the house, a part of the architectural scheme, as in the original Italian pergolas, or it may be half-hidden at the end of a garden or creeping along the edge of the woods. It may convert a path into a cloister or a grape arbor into a summer house. It has many traditions but no formal rules.

It has been used as a triumphant architectural feature in a modern country house;

on the foliage, pillars, furniture and floor.

As we have already said, in construction a pergola may relate closely to the architecture of the house, or on the other hand it may suggest an ornamental addition of a later date and be developed in materials different from the house, or it may bear no relation whatever to the house construction. The adobe pergola is a fascinating feature of many of the Pacific Slope houses; yet one often sees the adobe house with a pergola or pergola porch of redwood, designed on straight lines with Japanese effect. In New England and on Long Island the pergola with brick supports and wooden overhead beams is most usual, while out in New Jersey more often you find the pergola used in place of a porch, possibly a new feature of a quaint old house, and built of ordinary lumber, just as one would construct a trellis or a fruit arbor.

As a matter of fact, a pergola attached to the house is an ideal substitute for a piazza. This is especially true where there is the slightest tendency for the rooms to be somewhat dark, as it affords a decorative finish to the house, a charming resting place, a picturesque opportunity for vines, and yet permits all possible sunlight to reach the windows.

In one of the illustrations in this article the cement supports of the pergola are topped with rustic poles heavily draped with vines, and the effect is most picturesque. In fact, an entire rustic pergola is charming in an informal simple garden. It has, however, the drawback of not being as free from insects and dampness as the concrete structures.

As for the pergola "drapery," there is seemingly no limit to the beautiful things which the concrete or stone or brick columns will support. In the Far West some of the most beautiful pergolas are almost bowers of tea roses, intertwined with wistaria and monthly honeysuckle. In the East it is necessary to use the hardier roses, the Ramblers in different hues, white and red and pink. Wistaria is also one of the most attractive pergola vines when combined with others of the more hardy foliage and later bloom. It is difficult to get the monthly blooming honeysuckle in the East, but it proves most graceful as a pergola covering where it can be secured. Through the North, such vines as ivy, clematis and woodbine are all satisfactory, and nothing is more delightful than a pergola covered with grapevines, where the location and latitude are suitable, for the bloom of the grape is ineffable in the spring, the foliage is heavy through the summer and the fragrance and color of the fruit delicious in the fall. The delicately leaved jessamine with its sweet blossoms, the Allegheny vine, even more lace-like in foliage and graced by bells of white, the canary vine of yellow-orchid beauty, are unequaled for small, slender pergolas. The wild cucumber should be better known and appreciated, and also the bittersweet with its clusters of sweet white blossoms of springlike beauty, and its orange berries that break asunder at the first frost and reveal scarlet fruit which hang together, orange and scarlet, even when snow outlines twig and branch. The hop with its pale green pendant seed pods should be more in evidence in our garden as a decorative vine. The ornamental gourd is a quick-growing vine that can flourish verdantly while other vines are starting their slower climb, and its strange fruit can be put to a number of charming uses. The Dutchman's pipe is a vine whose curious flowers will repay cultivation.

It is always wise, in planning your garden, to plant about a pergola from two to four kinds of flower-bearing or fruit-bearing vines, so that each season will have its fragrance and color. It is also interesting to plant rows of shrubs at the foot of the supports and between the supports, that the whole structure may be more intimately connected with the ground.

Some pergolas are completely hidden by vines festooned from pillar to pillar; this is especially satisfactory in very hot climates. While others have vines twining only about the pillars with adequate protection overhead. This is by far the more classical and intrinsically beautiful method of treating a pergola. It has the disadvantage, however, of leaving the inner portion of the pergola a little less restful and homelike than when curtained by vines and shrubs.

For the newly built pergola there are many quick-growing vines which will give it a green and cheerful effect the first season,—morning-glo-

PERGOLA OF COBBLESTONES AND RUSTIC, WITH WISTARIA VINES.

A PERGOLA-ARBOR, SHOWING AN INTERESTING APPLICATION OF THE PERGOLA IDEA TO AN OLD-FASHIONED GRAPE ARBOR, ESPECIALLY ADAPTED TO THE MORE SIMPLE TYPE OF COUNTRY ARCHITECTURE.

ries, scarlet runners, clematis, with castor beans at the entrance and geraniums at the sides and you have by July the effect of many years' growth.

Pergola gateways are attractive when bowered by flowering vines, and a driveway arched at the entrance with a simple pergola has charm hard to excel. A division or retaining wall can be redeemed from monotony by using it as one side of a pergola, constructing the pillars of brick if the wall is of brick, or of stone or concrete if the wall is of either of these materials. The rafters can be of rustic or square-hewn beams. Such treatment of a wall would have quite the spirit of cloister walks, and seats built in would heighten this monastic quality.

As to the materials to be used in construction of all pergolas, the resources of the immediate locality should be drawn upon in preference to all others. Stone piers built of cobble will be most suitable to one neighborhood, while split stone is better in another, and in some places it would be possible to have them of whole field stone. Pillars of rough brick are decoratively valuable at times, terra cotta at others, cement at still others. They can be placed singly, in pairs or in groups to harmonize with the surrounding type of garden and house.

Turned wooden columns of classic design, either plain or fluted, are favorite supports for trellised roofs. Rustic pillars of cedar, fir, white pine, cypress, oak, madrone, redwood, with girders of the same wood a trifle small in size, are unequalled for informal gardens. Rustic is the most inexpensive material of which a pergola can be built, if it can be obtained with little cost of transportation, the square wooden supports coming next in order. Satisfactory combinations are sometimes devised, such as cement pillars and eucalyptus rafters and girders, stone supports with wooden rafters and trellis of various woods.

To preserve the true pergola form, to keep it from becoming an arbor, the trellis strips must not be put on horizontally between the pillars—this is the chief distinguishing note and must not be transgressed. Vines may be draped from pillar to pillar and not mar the purity of type, or trellis strips may be placed against the pillars, parallel with them, for vines to clamber upon, and purity of style be intact, but the horizontal feature must not appear upon the pergola—unless you want an arbor.

It is an excellent idea to plan a pergola with built-in seats at the sides and with rustic permanent tables, also with rough flooring for damp days. The joy of this garden feature is its livableness, to get the full satisfaction of which it must be a convenient homelike place for reading, sewing, afternoon tea, children's games. And of all things it should be the ideal spot for the writer or for the student, for working out of doors means working with health, and as a health-giving feature the properly constructed, properly draped pergola is second only to that other most wholesome development of modern building, the outdoor sleeping porch.

The pole pergola is a sort of pergola that is especially adapted to rustic surroundings, and many a restored cottage on abandoned farms has been made lovely by the introduction of such a feature, the poles having been cut from woodlot saplings. Even where all the materials had to be purchased,—cedar posts, plants, and the labor counted in—twelve or fifteen dollars, depending upon locality, would be fully sufficient to cover the whole cost.

The pergola-arbor illustrated on page one hundred and eighty-one is from the cottage of Mr. Frederick C. Keppel at Montrose, New York, designed by Edward Shepard Hewitt. For a pergola-arbor of this sort there could be no lovelier covering than the wild-grape, or the wild clematis, and, again, the kudzu vine, which comes to us from Japan and is found to be perfectly hardy everywhere, will, by reason of its extraordinarily rapid growth and luxuriant foliage of enormous rich green leaves, prove especially useful where a quick effect is desired.

The illustration on page one hundred and seventy-nine of a pergola porch on a country house at East Hampton, Long Island, exhibits another form of the pergola which requires far more restraint in planting, for it is intended that it should stand forth itself as an architectural feature; hence the vine-growth here will never be permitted completely to obscure the design of its support. The two great jars of terra-cotta add striking notes to the pergola and make this, in design, a successful house approach.

The pergola illustrated on page one hundred and seventy-eight is one connected with the outbuildings on the Stetson Farms, Sterlington, New York, designed by Alfred Hopkins. Here has been presented the problem of making the pergola serve, not only as a screen, but as a support for an overhead cartage rail which serves to facilitate the removal of stable litter expeditiously, neatly and hidden from observation. Ultimately the planting here will form a complete screen, summer and winter.

A PLEASANT VISTA OF APPROACH FORMED BY PERGOLA WITH RUSTIC ROOF SUPPORTED BY PILLARS OF CEMENT.

THE EFFECTIVE USE OF COBBLESTONES AS A LINK BETWEEN HOUSE AND LANDSCAPE

IN the building of modern country homes there seems to be no end to the adaptability of cobblestones and boulders in connection with the sturdier kinds of building material, for, if rightly placed with regard to the structure and the surroundings, they can be brought into harmony with nearly every style of architecture that has about it any semblance of ruggedness, especially if the surrounding country be hilly and uneven in contour and blessed—or cursed—with a plentiful crop of stones.

effect of a loose pile of stones. Very few houses that are possible for modern civilized life,—outside of the mountain camp—are sufficiently rough and primitive in construction to be exactly in harmony with the use of cobbles, and always there is a slight sense of effort when they are brought into close relation with finished structure.

Nevertheless the popularity of cobblestones and boulders for foundations, pillars, chimneys and even for such interior use as chimney-pieces, is unquestioned and in many cases the

Published in The Craftsman, November, 1908, Hunt & Eager, Architects.

CEMENT PAVED TERRACE OF A CALIFORNIA HOUSE, SHOWING EFFECT OF COBBLESTONES IN WALLS AND PILLARS, AND THE WAY THEY HARMONIZE WITH THE ROUGH SHINGLE AND TIMBER CONSTRUCTION.

We have never specially advocated the use of cobblestones in the building of Craftsman houses, for as a rule we have found that the best effects from a structural point of view can be obtained by using the split stones instead of the smaller round cobbles. Splitting the stone brings into prominence all the interesting colors that are to be found in field rubble and it is astonishing what a variety and richness of coloring is revealed when the stone is split apart so that the inner markings appear. Also a better structural line can be obtained when foundation and pillars are clearly defined instead of having somewhat the

effect is very interesting. There is growing up in this country, especially on the Pacific Coast, a style of house that seems to come naturally into harmony with this sort of stone work, and there is no denying that when the big rough stones and cobbles are used with taste and discrimination, they not only give great interest to the construction, but serve to connect the building very closely with the surrounding landscape.

The fact that we have found the best examples of this natural use of boulders and cobbles in California seems to be due largely to the influence of Japanese architecture over

THE EFFECTIVE USE OF COBBLESTONES

Published in The Craftsman, November, 1908, Hunt & Eager, Architects.

VERANDA AND TERRACE OF THE SAME HOUSE, GIVING A GOOD IDEA OF THE WAY COBBLESTONES MAY SERVE TO LINK THE HOUSE WITH THE SURROUNDING LANDSCAPE. THE EFFECT OF THE RUGGED FORM OF PILLARS AND PARAPET IN CONNECTION WITH THE SHINGLES OF THE WALL AND THE LACY FOLIAGE OF THE TREE IS ESPECIALLY STRIKING.

Published in The Craftsman, July, 1907, Greene & Greene, Architects.

A CALIFORNIA HOUSE MODELED AFTER THE JAPANESE STYLE, WITH HIGH RE-
TAINING WALL IN WHICH THE USE OF COBBLESTONES HAS PROVEN ESPECIALLY
DECORATIVE.

danger of incongruity, and on the other hand the stone is usually employed in a way that brings the entire building into the closest relationship with its environment.

The cobblestones used for the houses of this kind are of varying sizes. To give the best effect they should be neither too small nor too large. Stones ranging from two and one half inches in diameter for the minimum size to six or seven inches in diameter for the maximum size are found to be most generally suitable. Such stones, which belong of course to the limestone variety, and are irregularly rounded, can usually be obtained without trouble in almost any locality where there are any stones at all, picked up from rocky pasture land or a dry creek bottom. The tendency of builders is to select the whitest stones and the most nearly round that are obtainable.

This, however, applies only to the regular cobblestone construction as we know it in the East. In California the designers are much more daring, for they are fond of using large mossy boulders in connection with both brick and cobbles. The effect of this is singularly interesting both in color and form, for the warm purplish brown of the brick contrasts delightfully with the varying tones of the boulders covered with moss and lichen, and the soft natural grays and browns of the more or less primitive wood construction that is almost invariably used in connection with cobbles gives the general effect of a structure that

the new building art that is developing so rapidly in the West. In these buildings the use of stone in this form is as inevitable in its fitness as the grouping of rocks in a Japanese garden, for on the one hand the construction of the house itself is usually of a character that permits such a use of stone without

143

THE EFFECTIVE USE OF COBBLESTONES

PERGOLA, PORCH AND ENTRANCE OF A COUNTRY HOUSE AT RIDGEFIELD, CONNECTICUT. THE FOUNDATION AND FIRST STORY OF THE HOUSE ARE OF FIELD RUBBLE SET IN CEMENT, AND THE SECOND STORY IS BUILT OF OVER-BURNED BRICK WITH HALF-TIMBER CONSTRUCTION, GIVING A DELIGHTFUL COLOR EFFECT. THE HOUSE AND GARDEN ARE SO LINKED TOGETHER THAT THE FIRST IMPRESSION IS THAT OF PERFECT HARMONY AND CLOSE RELATIONSHIP, AN IMPRESSION THAT IS GREATLY HEIGHTENED BY THE USE MADE OF THE LOCAL STONE.

THE EFFECTIVE USE OF COBBLESTONES

Published in The Craftsman, July, 1907, Greene & Greene, Architects.

CONSTRUCTION OF THE PERGOLA AND ESPLANADE LEADING TO THE ENTRANCE OF A CALIFORNIA HOUSE. NOTE THE COMBINATION OF LARGE MOSSY BOULDERS WITH HARD-BURNED CLINKER BRICK SET IRREGULARLY IN DARK MORTAR.

in the illustration on page 105, where hard-burned brick and natural wood are most effectively combined with big rugged boulders and the large round slabs of stone that serve as steps. These stones, by their very conformation, p r o c l a i m themselves as belonging to New England, and the manner in which they are used is as definitely Eastern as the construction of the California houses is Western.

The Western method is admirably illustrated in the three different views given of the California house that so strongly reflects the influence of Japanese architecture. Here, instead of sharp-edged granite, we have big comfortable looking boulders with all the edges and corners worn off during the ages when they have rolled about in the mountain torrents, and the way they are w e d g e d helter-skelter among the irregular, roughly laid bricks of the walls, pillars and chimneys is as far from the conventional use of stone as is a Japanese garden from our own trim walks and flower beds. Such a combination as in shown in these pictures almost demands the suggestion of Japanese architecture in the house itself, and yet the whole thing belongs entirely to California.

has almost grown up out of the ground, so perfectly does it sink into the landscape around it.

The same effect is being sought more and more in the East by certain daring and progressive architects who, without regard to style and precedent, are building houses suited to the climate, the soil and the needs of life in this country. An excellent example of this is shown

The harmony of this house with its surroundings will be understood when we say that it is situated on high ground overlooking the wild gorge of the Arroyo Seco and that the trees close to it are gnarled, hoary oaks, towering eucalyptus, widespreading cottonwoods, tall, slim poplars and sycamores.

THE EFFECTIVE USE OF COBBLESTONES

Published in The Craftsman, November, 1907.

A HOUSE NEAR PASADENA, CALIFORNIA, SHOWING THE STRIKING EFFECT GAINED BY THE USE OF COBBLESTONES AND BOULDERS IN THE FOUNDATION, CHIMNEY AND YARD WALL.

Published in The Craftsman, November, 1907.

A CALIFORNIA HOUSE WHERE THE USE OF COBBLESTONES IN THE STRUCTURE ITSELF IS REPEATED IN THE LOW PILLARS THAT MARK THE ENTRANCE OF WALK AND DRIVEWAY AND IN THE GARDEN WALL, THUS DRAWING CLOSER THE RELATIONSHIP BETWEEN HOUSE AND GROUND.

THE EFFECTIVE USE OF COBBLESTONES

Published in The Craftsman, July, 1907, Greene & Greene, Architects.

A HOUSE IN SOUTHERN CALIFORNIA THAT SHOWS STRONG TRACES OF JAPANESE INFLUENCE, AS EVIDENT IN THE USE OF COBBLESTONES AND BOULDERS IN COMBINATION WITH BRICK, AS IN THE STRUCTURE ITSELF.

Published in The Craftsman, July, 1907, Greene & Greene, Architects.

AN EXCELLENT EXAMPLE OF THE RIGHT USE OF COBBLES AND BOULDERS. ANOTHER VIEW OF THE SAME HOUSE, SHOWING THE WAY IN WHICH THE GRACEFUL LINE OF THE CHIMNEY RISES FROM THE WALL OF BRICK AND STONE, AND ALSO THE MANNER IN WHICH THE STONE IS CARRIED PART WAY UP THE CHIMNEY SO THAT IT SHOWS IRREGULARLY HERE AND THERE.

BEAUTIFUL GARDEN GATES: THE CHARM THAT IS ALWAYS FOUND IN AN INTERESTING APPROACH TO AN ENCLOSURE

FEW people realize how much depends upon the approach to any given place. A pleasant entrance that rouses the interest and conveys some impression of individuality seems an earnest of pleasant things to come and is always associated in the memory with the anticipation that came from that first impression. Especially is this true of a garden gate, which for most of us holds a suggestion of sentiment and poetry because it is in its own way a symbol; it leads out to greater spaces or inward to more intimate beauty. Even to the most prosaic it always holds something of a promise of the peaceful and pleasant place that lies within. Thus it seems right that a garden gate should have a charm and grace all its own; that it should be embowered with trailing vines and blooming flowers in summer time and should always hold forth the inviting suggestion of pleasure and welcome beyond.

The illustrations given here are all of very simple garden gateways that are made attractive by the method of construction, by the placing of vines and flowers or by some graceful conceit in outline and relation to the surroundings. The hooded gate shown on this page forms a charming link between garden and garden. One may rest a moment within its shade and it seems to bind together the two plots of green divided by the fence. The trellised arbor and archway which spans the flower walk in an English garden is illustrated here because of the charming suggestion it contains for making a division between two parts of the same garden. The "pergola gate" shown below is illustrated without the vines that are meant to clothe it, because we desire to give a clear idea of the construction. The finely planned proportions of the heavy timbers and the straight unornamented lines suggest an inspiration from Japan. The vine covered rustic arbor which arches over the walk leading to the entrance of the house beyond is hardly a garden gate, yet it comes within the same class because it furnishes a most attractive approach to house and garden.

Published in The Craftsman, June, 1908.

Courtesy of John Lane Company.

ARBOR AND FLOWER WALK IN AN ENGLISH GARDEN, AFFORDING NOT ONLY A PLEASANT SUMMER RETREAT BUT ALSO A MOST ATTRACTIVE VISTA THROUGH THE LARGE GROUNDS.

Published in The Craftsman, June, 1908.

GATE WITH PERGOLA CONSTRUCTION OVERHEAD MEANT TO SERVE AS A SUPPORT FOR CLIMBING VINES.

HOMEMADE RUSTIC ARBOR, COVERED WITH CLIMBING ROSES AND HONEYSUCKLE, PLACED AT THE ENTRANCE OF THE WALK LEADING TO THE FRONT DOOR. ONE SUCH STRUCTURAL FEATURE AS THIS WOULD SERVE AS THE CENTRAL POINT OF INTEREST IN AN ENTIRE GARDEN.

Published in The Craftsman, March, 1907.

A VERY SIMPLE RUSTIC GATEWAY MADE OF TWO UPRIGHTS WITH CROSS-PIECES THAT SERVE AS A SUPPORT FOR VINES, AND A PEAKED HOOD OF THE SAME CONSTRUCTION, THE FENCE AND GATE ARE ALSO OF RUSTIC CONSTRUCTION. SUCH AN ENTRANCE ADDS A TOUCH OF DIGNITY AS WELL AS PICTURESQUENESS TO THE SIMPLEST GARDEN.

THE NATURAL GARDEN: SOME THINGS THAT CAN BE DONE WHEN NATURE IS FOLLOWED INSTEAD OF THWARTED

MAKING a garden is not unlike building a home, because the first thing to be considered is the creation of that indefinable feeling of restfulness and harmony which alone makes for permanence. Therefore, in planning a garden that we mean to live with all our lives, it is best to let Nature alone just as far as possible, following her suggestions and helping her to carry out her plans by adjusting our own to them, rather than attempting to introduce a conventional element into the landscape.

We have already explained in detail the importance of building a house so that it becomes a part of its natural surroundings; of planning it so that its form harmonizes with the general contour of the site upon which it stands and also of the surrounding country, and of using local materials and natural colors, wherever it is possible, so that the house may be brought into the closest relationship with its natural surroundings. But no matter how well planned the house may be, or how completely in keeping with the country, the climate and the

life that is to be lived in it, the whole sense of home peace and comfort is gone if the garden is left to the mercy of the average gardener, whose chief ambition usually is to achieve trim walks, faultless flower-beds and neatly barbered shrubs, and whose appreciation of wild natural beauty is small.

To give a real sense of peace and satisfaction a garden must be a place in which we can wander and lounge, pick flowers at our will and invite our souls, and we can do none of these if we have the feeling that trees, shrubs and flowers were put there arbitrarily and according to a set, artificial pattern, instead of being allowed to grow up as Nature meant them to do. Therefore, knowing the vital importance of the right kind of garden to the general scheme, we have given here some examples of the natural treatment of moderate-sized grounds, trusting that they may be suggestive to home builders. The house shown in the illustrations was built by an artist out in a pasture lot and the garden that has been encouraged to grow up around it has more of the

Published in The Craftsman, January, 1908.

A FLIGHT OF STEPS WHICH HAVE BEEN CUT OUT FROM THE SIDE OF A HILL AND REINFORCED WITH HEAVY BOARDS ROUNDED AT THE EDGE. THE CURVING LINE OF THE STEPS, WHICH CONFORMS TO THE CONTOUR OF THE HILL, AND THE DRAPERY OF VINES AND NATURAL UNDERGROWTH THAT COVERS THE RUSTIC RAILING ON EITHER SIDE GIVES TO THIS APPROACH A RARE AND COMPELLING CHARM.

AN EXAMPLE OF THE EFFECT PRODUCED BY THE LAVISH USE OF VINES UPON A HOUSE WHERE THEY NATURALLY BELONG. THE CONSTRUCTION OF COBBLESTONE AND ROUGH CEMENT SEEMS TO DEMAND JUST SUCH GRACIOUS DRAPERY TO BRING IT INTO STILL CLOSER RELATIONSHIP WITH ITS SURROUNDINGS.

154

feeling of free woods and meadows than of a primly kept enclosure. The trees were thinned out just enough to allow plenty of air and sunshine and the sense of space that is so necessary, and, for the rest, were permitted to grow as they would. As Nature never makes a mistake in her groupings, the different varieties of trees fall into the picture in a way that could never be achieved by the most ingenious planting. Such shrubs and flowers as have been set out are of the more hardy varieties that belong to the climate and to the soil, and the vines that clamber over the low stone garden walls and curtain the walls of the house seem more to belong to the wild growths of the hillside than to have been planted by man. Where there is a path or a flight of steps the course of it is ruled by the contour of the ground so that the whole impression is that of Nature smoothed down in places and in others encouraged to do her very best.

These pictures, of course, are only suggestive, for in the very nature of things this kind of a garden cannot be made by rule, as no two places require or will admit the same treatment. The only way to obtain the effect desired is to cultivate the feeling of kinship with the open country and with growing things, and so to learn gradually to perceive the orig-

inal plan. After that, all that is needed is to let things alone so far as arrangement goes, and to work in harmony with the thing that already exists.

Most fortunate is the home builder who can set his house out in the open where there is plenty of meadowland around it and an abundance of trees. If the ground happens to be uneven and hilly, so much the better, for the gardener has then the best of all possible foundations to start from and, if he be wise, he will leave it much as it is, clearing out a little here and there, planting such flowers and shrubs as seem to belong to the picture and allowing the paths to take the directions that would naturally be given to footpaths across the meadows or through the woods,—paths which invariably follow the line of the least resistance and so adapt themselves perfectly to the contour of the ground.

In connection with these garden pictures we give several illustrations of the effect of an abundant growth of vines over the walls of the house and around its foundations, and also show in one picture the result that can be obtained by allowing a fast growing vine to form a leafy shade to the porch that is used as an outdoor living room. The lattice construction of the roof admits plenty of sunlight.

Published in The Craftsman, December, 1907.

VINE COVERED PORCH THAT IS USED AS AN OUTDOOR LIVING ROOM AND THAT SEEMS MORE A PART OF THE GARDEN THAN OF THE HOUSE.

WHAT MAY BE DONE WITH WATER AND ROCKS IN A LITTLE GARDEN

WE have to acknowledge our indebtedness to the Japanese for more inspiration in matters of art and architecture than most of us can realize, and in no department of art is the realization of subtle beauty that lies in simple and unobtrusive things more valuable to us as home makers than the suggestions they give us as to the arrangement of our gardens. With our national impulsiveness, we are too apt to go a step beyond the inspiration and attempt direct imitation, which is a pity, because the inevitable failure that must necessarily attend such mistaken efforts will do more than anything else to discourage people with the idea of trying to have a Japanese garden. But if we once get the idea into our heads that the secret of the whole thing lies in the exquisite sense of proportion that enables a Japanese to produce the effect of a whole landscape within the compass of a small yard, there is some hope of our being able to do the same thing in our own country and in our own way.

Our idea of a garden usually includes a profusion of flowers and ambitious-looking shrubs, but the Japanese is less obvious. He loves flowers and has many of them, but the typical Japanese garden is made up chiefly of stones, ferns, dwarf trees and above all water. It may be only a little water,—a tiny, trickling stream not so large as that which would flow from a small garden hose. But, given this little stream, the Japanese gardener,—or the American gardener who once grasps the Japanese idea,—can do wonders. He can take that little stream, which represents an amount of water costing at the outside about three dollars a month, and can so direct it that it pours over piles of rocks in tiny cascades, forming pool after pool, and finally shaping its course through a miniature river into a clear little lake. If it is a strictly Japanese garden, both river and lake will be bridged and the stream will have as many windings as possible, to give a chance for a number of bridges. Also it will have temple lanterns of stone, bronze storks and perhaps a tiny image of Buddha.

But in the American garden we need none of these things, unless indeed we have space enough so that a portion of the grounds may be devoted to a genuine Japanese garden like the one shown in the illustrations. This indeed might have been picked up in Japan and transplanted bodily to America, for it is the garden of Mr. John S. Bradstreet, of Minneapolis, who is a lover of all things Japanese and has been in Japan many times. This garden occupies a space little more than one hundred feet in diameter, and yet the two illustrations we give are only glimpses of its varied charm. They are chosen chiefly because they illustrate the use that can be made of a small stream of water so placed that it trickles over a pile of rocks. The effect produced is that of a mountain glen, and so perfect are the proportions and so harmonious the arrangement that there is no sense of incongruity in the fact that the whole thing is on such a small scale.

Where people have only a small garden, say in the back yard of a city home or in some nook that can be spared from the front lawn, an experiment with the possibilities of rocks, ferns and a small stream of water would bring rich returns. We need no temple lanterns or images of Buddha in this country, but we do need the kind of garden that brings to our minds the recollection of mountain brooks, wooded ravines and still lakes, and while it takes much thought, care and training of one's power of observation and adjustment to get it, the question of space is not one that has to be considered, and the expense is almost nothing at all.

The thing to be most avoided is imitation either of the Japanese models from which we take the suggestion for our own little gardens or of the scenery of which they are intended to remind us. It is safest to regard such gardens merely as an endeavor on our part to create something that will call into life the emotion or memory we wish to perpetuate.

All these suggestions are for a small garden such as would naturally belong to a city or suburban house, but if such effects can be produced here in a corner and by artificial means, it is easy to imagine what could be done with large and naturally irregular grounds, say on a hillside, or where a natural brook wound its way through the garden, giving every opportunity for the picturesque effects that could be created by very simple treatment of the banks, by a bridge or a pool here and there and by a little adjustment of the rocks lying around.

Courtesy of Country Life in America.

A PART OF A JAPANESE GARDEN OWNED BY MR. JOHN S. BRADSTREET, OF MINNEAPOLIS.   AN EXCELLENT EXAMPLE OF HOW ROCKS, DWARF TREES AND A TINY STREAM OF WATER MAY BE USED TO MAKE A HIGHLY DECORATIVE EFFECT.

ANOTHER PART OF MR. BRADSTREET'S GARDEN, SHOWING BRIDGES MADE OF WATER-WORN TEAKWOOD TAKEN FROM AN OLD JUNK. THE FOUNTAIN, PILE OF ROCKS AND DWARF TREES ARE SEEN FROM A DIFFERENT ANGLE.

EXAMPLE OF WHAT MAY BE DONE WITH A VERY SMALL SUPPLY OF WATER. THE POOL HERE IS FED SOLELY BY A TINY STREAM WHICH ISSUES FROM THE DRAGON'S MOUTH AND FORMS A SLENDER CASCADE OVER THE ROCKS.

REST HOUSE AND POOL IN MR. BRADSTREET'S JAPANESE GARDEN, SHOWING TEMPLE LANTERN, SMALL IMAGE **OF** BUDDHA AND THE EFFECT OF ROCKS AROUND THE MARGIN FRAMING THE AQUATIC PLANTS IN THE POOL.

HALLS AND STAIRWAYS: THEIR IMPORTANCE IN THE GENERAL SCHEME OF A CRAFTSMAN HOUSE

WITH the general adoption of the simpler and more sensible ideas of house building that have come to the front in late years, the hall seems to be returning to its old-time dignity as one of the important rooms of the house. Instead of the small dark passageway, with just room enough for the hat tree and the stairs, that we have long been familiar with in American houses, we have now the large reception hall with its welcoming fireplace and comfortable furnishings,—as inviting a room as any in the house. There is even a suggestion of the "great hall of the castle," where in bygone days all indoor life centered, in the ever-increasing popularity of the plan which throws hall, living room and dining room into one large irregular room, divided only by the decorative post-and-panel construction that we so frequently use to indicate a partition, or by large screens that serve temporarily to shut off one part or another if privacy should be required. In this main room all guests are received, all the meals are served and the greater part of the family life is carried on. Even where this plan is not adopted and the rooms of the lower story are completely separated from one another, the large reception hall is still counted as one of the principal rooms of the house, and what used to be considered the entrance or stair hall is now either absent entirely or treated as a vestibule; generally curtained off from the reception hall or living room into which it opens in order to prevent drafts from the entrance door.

Whether it be a large or small reception hall, or an entrance only large enough for the stairs and the passageway from the front door to the other rooms in the house, the hall is always worthy of careful consideration as to structural features and color scheme, for it gives the first impression of the whole house. It is the preface to all the rest and in a well planned house it strikes the keynote of the whole scheme of interior decoration. Above all things, the hall ought to convey the suggestion of welcome and repose. In a cold climate, or if placed on the shady side of the house, it is worth any pains to have the hall well lighted and airy and the color scheme rich and warm. It is the first impression of a house that influences the visitor and the sight of a cheerless vista upon entering chills any appreciation of subsequent effects. With a sunny exposure, or in a country where heat has to be reckoned with for the greater part

Published in The Craftsman, January, 1906.

A TYPICAL CRAFTSMAN STAIRWAY WITH LANDING USED AS A STRUCTURAL FEATURE OF THE RECEPTION HALL. THIS IS AN EXCELLENT EXAMPLE OF THE POST-AND-PANEL CONSTRUCTION WHICH IS SO OFTEN USED TO INDICATE THE DIVISION BETWEEN TWO ROOMS.

Published in The Craftsman, November, 1906.

AN UPPER HALL WHICH IS FITTED UP FOR USE AS A SEWING ROOM, STUDY, OR PLAYROOM, ACCORDING TO THE USE FOR WHICH IT IS MOST NEEDED. SUCH AN UPSTAIRS RETREAT IS DELIGHTFUL IN A HOUSE WHERE THE ARRANGEMENT OF THE WHOLE LOWER STORY IS OPEN, AS IT AFFORDS A MORE OR LESS SECLUDED PLACE FOR WORK OR STUDY AND YET HAS THE FREEDOM AND AIRINESS OF A LARGE SPACE.

of the year rather than cold, an effect of restful shadiness and coolness is quite as inviting in its way, although it is always safe to avoid a cold color scheme for a hall, as the suggestion it conveys is invariably repellent rather than welcoming.

In England the large hall designed for the general gathering place of the family is a feature in nearly every moderately large house, particularly in the country. These English halls are always roomy and comfortable and in many cases are both picturesque and sumptuous in effect, having a certain rich stateliness that seems to have descended in direct line from the great hall of old baronial days. In this country the hall is more apt to be a part of the living room, and, while quite as homelike and inviting, is simpler in style.

The illustration on page 125 shows the part of a Craftsman reception hall that contains the stairway. A small den or lounging room is formed by the deep recess that appears at one side of the staircase, which is central in position and is completely masked, excepting the lower steps and the landing, by the post construction above the solid wainscot that surrounds it. This wainscot turns outward to the width of a single panel at either side of the stair, one sheltering the end of the seat built in at the right side and the other partially dividing off the recess to the left. So arranged, the staircase forms an important part of the decorative treatment of the room.

The second illustration shows an upstairs hall, which has somewhat the effect of a gallery, as it is open to the stairway except for a low balustrade. This nook in the upper hall takes the place of a sewing room or an upstairs sitting room, and is infinitely more attractive because of the freedom and openness of the arrangement. While not in any sense a separate room, it still allows a certain seclusion

Published in The Craftsman, January, 1906.

RECEPTION HALL AND STAIRCASE WHERE THE LANDING PROJECTS INTO THE ROOM ALMOST DIRECTLY OPPOSITE THE ENTRANCE DOOR. THIS HALL IN MOST CRAFTSMAN HOUSES IS LITTLE MORE THAN A NOOK IN THE LIVING ROOM.

for anyone who wishes to read, work or study.

The third illustration shows another Craftsman reception hall in which the staircase is the prominent structural feature. The double casements light stair and landing and also add considerably to the light in the room. Just below the stair is a comfortable seat with the radiator hidden below, and a coat closet fills the space between the seat and the wall.

A larger hall that is emphatically a part of the living room is seen in the last illustration. Here there is no vestibule and the wide entrance door with the small square panes in the upper part belong to the structural decoration of the room. Additional light is given from the same side by the row of casements recessed to leave a wide ledge for plants. The ceiling is beamed and the whole construction of the room is satisfying, although interest at once centers upon the staircase as the prominent structural feature. This is in the center of the room and has a large square landing approached by three shallow steps. The stairs run up toward the right at the turn and the space between steps and ceiling is filled with slim square uprights, two on each step, which give the effect of a grille, very open and very decorative. Opposite the stair on the landing is a railing about the height of a wainscot with posts above. Treated in this manner, the staircase seems intended as much for beauty as for utility, and so fulfills its manifest destiny in the Craftsman decorative scheme.

In a small house there are often many considerations which prevent the use of the hall as a living room. Many people object to the draughts and waste of heat entailed by the open stairway and prefer a living room quite separate from the entrance to the house. In this case it is better to omit the reception hall and to have merely a small entrance hall, rather than the compromise that contains no possibility of comfort and yet is crammed with all the features that belong in the larger hall intended for general use. An entrance hall of this kind may be made very attractive and inviting by the wise selection of the woodwork and color scheme and by care in the designing of the stairway, which of course is the principal structural feature in any hall.

THE LIVING ROOM: ITS MANY USES AND THE POSSIBILITIES IT HAS FOR COMFORT AND BEAUTY

UNQUESTIONABLY the most important room in the house is the living room, and in a small or medium sized dwelling this room, with the addition of a small hall or vestibule and a well-planned kitchen, is all that is needed on the first floor. A large and simply furnished living room, where the business of home life may be carried on freely and with pleasure, may well occupy all the space that is ordinarily partitioned into small rooms conventionally planned to meet supposed requirements. It is the executive chamber of the household, where the family life centers and from which radiates that indefinable home influence that shapes at last the character of the nation and the age. In the living room of the home, more than in almost any other place, is felt the influence of material things. It is a place where work is to be done and it is also the haven of rest for the worker. It is the place where children grow and thrive and gain their first impressions of life and of the world. It is the place to which a man comes home when his day's work is done and where he expects to find himself comfortable and at ease in surroundings that are in harmony with his daily life, thought and pursuits.

In creating a home atmosphere, the thing that pays and pays well is honesty. A house should be the outward and visible expression of the life, work and thought of its inmates. In its planning and furnishing, the station in life of its owner should be expressed in a

CHIMNEYPIECE AND FIRESIDE SEATS IN A TYPICAL CRAFTSMAN LIVING ROOM. THE CHIMNEYPIECE IS PANELED WITH DULL-FINISHED GRUEBY TILES BANDED WITH WROUGHT IRON HELD IN PLACE BY COPPER RIVETS. THE FIRE-PLACE HOOD IS OF COPPER AND THE PANELING OF SEATS AND WAINSCOT IS IN FUMED OAK.

Published in The Craftsman, December, 1905.

A LIVING ROOM WHICH IS ALSO USED FOR LIBRARY AND WORK ROOM. NOTE THE WAY IN WHICH THE DESK WITH ITS DRAWERS AND PIGEON HOLES IS BUILT INTO THE WALL SO THAT IT FORMS A PART OF THE BOOKCASE. THE LONG ROW OF CASEMENTS WITH THE WINDOW SEAT BELOW NOT ONLY FLOODS THE ROOM WITH LIGHT BUT FORMS A DECORATIVE FEATURE OF THE CONSTRUCTION.

THE LIVING ROOM

A FIRESIDE NOOK THAT IS DEEPLY RECESSED FROM THE LIVING ROOM. THE CEILING OF THE NOOK IS MUCH LOWER THAN THAT OF THE MAIN ROOM, GIVING AN EFFECT OF COMFORT THAT IS HARD TO OBTAIN IN ANY OTHER WAY.

dignified manner, not disguised. If servants cannot be afforded without too heavy a tax upon the family finances, build the house so that it is convenient to get along without them. It is astonishing how easy the care of a house can be made by the simple process of eliminating unnecessary things. The right kind of a home does not drag out all that there is in a man to keep it going, nor is the care of it too heavy a burden upon a woman. It should be so planned that it meets, in the most straightforward manner, the actual requirements of those who live in it, and so furnished that the work of keeping it in order is reduced to a minimum.

It is the first conception of a room that decides whether it is to be a failure or a success as a place to live in, for in this lies the character that is to be uniquely its own. In every house, however, modest, there can be a

CORNER OF A LIVING ROOM THAT IS ALSO USED AS A WORK ROOM. THE PANELING ON EITHER SIDE OF THE CHIM-NEYPIECE EXTENDS TO THE CEILING SO THAT THE ENTIRE WALL SPACE IS LINED WITH WOOD.

Published in The Craftsman, July, 1906.

CHIMNEYPIECE IN A MODERATE SIZED LIVING ROOM WHERE THE WALL TREATMENT ALLOWS THE INTRODUCTION OF A SHADOWY LANDSCAPE FRIEZE. THIS BEING SO DEFINITELY DECORATIVE, THE WALL SPACES BELOW ARE LEFT ABSOLUTELY PLAIN AND THE STRUCTURAL FEATURES ARE ALSO SEVERELY SIMPLE IN CHARACTER.

THE LIVING ROOM

living room that shows an individuality possessed by no other, an individuality that is actually a part of the place, if the room be planned to meet the real needs of those who are to live in it and to turn to the best advantage the conditions surrounding it. These conditions are as many as there are rooms. The situation and surroundings of the plot of ground on which a house is built has much to do with the position of the living room in the plan of that house. As it is the principal room, it should have an exposure which insures plenty of sunlight for the greater part of the day and also the pleasantest outlook possible to the situation. Both of these considerations, as well as the best arrangement of wall spaces, govern the placing of the windows and of outside doors, which may open into the veranda, the sun room, or the garden.

The structural variations of the living room are endless, as they are dominated by the tastes and needs of each separate family. If

Published in The Craftsman, February, 1902.

A RECESSED WINDOW SEAT THAT WOULD SERVE FOR ANY ROOM IN THE HOUSE.

the room is to be a permanently satisfying place to live in, nothing short of the exercise of individual thought and care in its arrangement will give the result. But one thing must be kept in mind if the room is to be satisfactory as a whole, and that is, to provide a central point of interest around which the entire place is built, decorated and furnished, for it gives the keynote both as to structure and color scheme. It may be a well planned fireplace, either recessed or built in the ordinary manner, with fireside seats, bookcases,

Published in The Craftsman, October, 1905.

A RECESSED FIREPLACE NOOK IN A ROOM WHERE THE WOODWORK IS LIGHT AND FINE AND THE PANELED WALL SPACES ARE COVERED WITH SOME FABRIC SUCH AS SILK, CANVAS, OR JAPANESE GRASS CLOTH.

THE LIVING ROOM

Published in The Craftsman, February, 1907.

A CHARACTERISTIC CRAFTSMAN INTERIOR, SHOWING THE ENTRANCE HALL, STAIRCASE AND LANDING AND A PART OF THE LIVING ROOM. NOTE THE WAY IN WHICH THE LINE OF THE MANTEL SHELF IS CARRIED THE WHOLE LENGTH OF THE WALL BY THE TOPS OF THE BUILT-IN BOOKCASES AND HOW IT IS FINISHED BY THE BALUSTRADE OF THE STAIR LANDING. ALSO NOTE THE MANNER IN WHICH THE ENTRANCE HALL IS DIVIDED FROM THE LIVING ROOM SO THAT ITS SEPARATENESS IS INDICATED WITHOUT DESTROYING THE SENSE OF SPACE WHICH MEANS SO MUCH TO THE BEAUTY OF THE MAIN ROOM.

Published in The Craftsman, November, 1905.

BUILT-IN CHINA CLOSETS ON EITHER SIDE OF THE FIREPLACE IN A LIVING ROOM WHICH IS ALSO USED AS A DINING ROOM. BY A SLIGHT DIFFERENCE IN ARRANGEMENT THE CUPBOARDS ABOVE COULD BE MADE TO SERVE AS BOOKCASES AND THOSE BELOW AS STORAGE PLACES FOR PAPERS, MAGAZINES AND THE LIKE.

THE LIVING ROOM

cupboards, shelves, or high casement windows so arranged as to be an integral part of the structure. The chimneypiece strikes a rich color-note with its bricks or tiles and glowing copper hood, and the woodwork, wall spaces and decorative scheme are naturally brought into harmony with it. Or perhaps the dominant feature may be the staircase, with its broad landing and well-designed balustrade; or it may be a group of windows so placed that it makes possible just the right arrangement of the wall spaces and commands the best of the view. Or if living room and dining room are practically one, the main point of interest may be a sideboard, either built into a recess or, with its cupboards on either side and a row of casement windows above, occupying the entire end of the room.

Any commanding feature in the structure of the room itself will naturally take its place as this center of interest; if there are several, the question of relative importance will be easily settled, for there can be only one dominant point in a well planned room. The English thoroughly understand the importance of this and the charm of their houses depends largely upon the skilful arrangement of interesting structural features around one center of attention to which everything else is subordinate. Also the English understand the charm of the recess in a large room. Their feeling regarding it is well expressed by a prominent English architect of the new school who writes: "Many people have a feeling that there is a certain cosiness in a small room entirely unattainable in a large room; this is a mistake altogether; quite the reverse has been my experience, which is, that such a sense

FIREPLACE IN A LIVING ROOM. THE SQUARE MASSIVE CHIMNEYPIECE IS BUILT OF HARD-BURNED RED BRICK LAID UP IN DARK MORTAR WITH WIDE JOINTS. THE MANTEL SHELF AS ILLUSTRATED HERE IS OF RED CEMENT, BUT A THICK OAK PLANK WOULD BE EQUALLY EFFECTIVE. THE HOOD IS OF COPPER AND THE FIREPLACE IS BANDED WITH WROUGHT IRON. THE PANELING ABOVE THE BOOKCASES GIVES AN INTERESTING DIVISION OF THE WALL SPACES.

THE LIVING ROOM

of cosiness as can be got in the recesses of a large room can never be attained in a small one. But if your big room is to be comfortable, it must have recesses. There is a great charm in a room broken up in plan, where that slight feeling of mystery is given to it which arises when you cannot see the whole room from any one place in which you are likely to sit; when there is always something around the corner."

Where it is possible, the structural features that actually exist in the framework should be shown and made ornamental, for a room that is structurally interesting and in which the woodwork and color scheme are good has a satisfying quality that is not dependent upon pictures or bric-a-brac and needs but little in the way of furnishings.

Only such furniture as is absolutely necessary should be permitted in such a room, and that should be simple in character and made to harmonize with the woodwork in color and finish. From first to last the room should be treated as a whole. Such furniture as is needed for constant use may be so placed that it leaves plenty of free space in the room and when once placed it should be left alone. Nothing so much disturbs the much desired home atmosphere as to make frequent changes in the disposition of the furniture so that the general aspect of the room is undergoing continual alteration. If the room is right in the first place, it cannot be as satisfactorily arranged in any other way. Everything in it should fall into place as if it had grown there before the room is pronounced complete.

Published in The Craftsman, January, 1906.

WINDOW SEAT IN A LIVING ROOM. THE GROUP OF WINDOWS WITH THE SEAT BELOW EXTENDS ACROSS THE ENTIRE END OF THE ROOM AND THE TWO ENDS OF THE SEAT ARE FORMED BY THE SMALL SQUARE BOOKCASES BUILT INTO THE CORNERS.

THE DINING ROOM AS A CENTER OF HOSPITALITY AND GOOD CHEER

NEXT to the living room the most important division of the lower floor of a house is the dining room. The living room is the gathering place of the household, the place for work as well as for pleasure and rest; but the dining room is the center of hospitality and good cheer, the place for in a carefully planned house the work of the household is made as easy as possible. Hence it goes without saying that the dining room should be placed in such relation to the kitchen that the work of serving meals goes on with no friction and with as few steps as possible. A noiseless and well fitted swing

Published in The Craftsman, January, 1906.

CRAFTSMAN DINING ROOM WITH SIDEBOARD BUILT INTO A RECESS AND SURMOUNTED BY FOUR CASEMENT WINDOWS. NOTE HOW THE GROUPING OF THE WINDOWS IS REPEATED AT THE END OF THE ROOM.

that should hold a special welcome for guests and home folk alike. Instead of being planned to fulfil manifold functions like the living room, it has one definite use and purpose and no disturbing element should be allowed to creep in.

In planning a dining room two considerations take equal rank,—convenience and cheerfulness. Convenience must come first, door serves as a complete bar against sounds and odors from the kitchen, even if the connection be direct. If a butler's pantry should be preferred for convenience in serving, it would naturally be placed between the kitchen and the dining room. Much time and many steps are saved also if the principal china cupboard is built in the wall between the dining room and kitchen or butler's pantry, with

173

Published in The Craftsman, October, 1905.

DINING ROOM WITH DISH CUPBOARD BUILT INTO THE WALL SO THAT THE DOORS ARE FLUSH WITH THE SURFACE
THE SIDEBOARD IN THIS CASE IS MOVABLE AND THE REMAINDER OF THE WALL SPACE BELOW THE FRIEZE IS TAKEN
UP BY A PICTURE WINDOW IN WHICH ARE ACCENTED THE COLORS THAT PREVAIL IN THE DECORATION OF THE ROOM.

Published in The Craftsman, November, 1905.

RECESSED WINDOW AND SEAT IN A DINING ROOM. AN UNUSUALLY QUAINT EFFECT IS GIVEN BY THE SMALL LEADED
PANES OF GLASS AND THE BROAD WINDOW LEDGE FOR HOLDING PLANTS.

THE DINING ROOM

SIDEBOARD BUILT INTO A RECESS WITH DISH CUPBOARDS ON EITHER SIDE. SQUARE MATT-FINISHED TILES ARE USED
TO FILL IN THE PANELS ABOVE THESE CUPBOARDS AND THE SPACE BETWEEN THE TOP OF THE SIDEBOARD AND THE
WINDOW LEDGE.

WINDOW EXTENDING THE WHOLE WIDTH OF A DINING ROOM AND INTENDED FOR AN EXPOSURE WHERE THERE IS
AN ESPECIALLY FINE VIEW.

doors opening on both sides so that dishes may be put away after washing without the necessity of carrying them into the dining room. Such an arrangement results in a great saving of broken china as well as in added convenience. This kind of a china cupboard may be made very decorative by putting small-paned or leaded glass doors on the dining room side and treating the wooden doors at the back like the wood trim of the room, which makes an effective setting for the china.

of cheerfulness may be given by the warmth of color in the room. A richness and decision of wall coloring that would grow wearisome in a room lived in all the time has all the pleasant and enlivening effects of a change when seen occasionally in a dining room. If the dining room is to be a part of the living room, it is well to plan it as one would a large recess. In that case the color scheme should, of course, be in close harmony with that of the living room; but even then it may strike a stronger

Published in The Craftsman, November, 1905.

ANOTHER FORM OF BUILT-IN SIDEBOARD WITH LINEN DRAWERS ON EITHER SIDE. THIS IS INTENDED TO FILL THE WHOLE SPACE ACROSS THE END OF THE DINING ROOM.

If possible, the dining room should have an exposure that gives it plenty of light as well as air. The windows play such an important part in the decoration of a room that a pleasant outlook is greatly to be desired. The brilliance of a sunny exposure may always be tempered by a cool and restful color scheme in walls and woodwork. On the other hand, if the room has a shady exposure and threatens to be somber on dark days, an atmosphere

and more vivid note in the walls, while the woodwork remains uniform throughout. A large screen placed in the opening of the recess may be made very decorative if it serve as a link in the color scheme as well as the leading element in that pleasant little sense of mystery that always accompanies a glimpse of something partially unseen.

Nowhere more than in the dining room is evidenced the value of structural features.

THE DINING ROOM

Almost all the decorative quality of the room depends upon them. In addition to wainscot and ceiling beams,—or instead of them if the room be differently planned,—the charm of well placed windows, large and small; of built-in cupboards, sideboards and cabinets for choice treasures of rare china or cut glass; of shelves and plate rack; of window ledge and window seat; and above all of a big cheery fireplace, is as never-ending as the ingenuity which gives to each really beautiful room exactly what it needs. And always it should be remembered that, in the dining room as in the living room, there should be one central structural feature which dominates all the rest.

Some examples of these ruling features are given in the accompanying illustrations. In one there is the wide sideboard built into a recess surmounted by three casement windows and flanked by a small china cupboard on either side. In another a wide window is recessed, giving a broad ledge for the growing things that always add beauty and life to a room. Still another recessed window shows a row of small-paned casements with plant ledge and a well cushioned seat below.

A GROUP OF CRAFTSMAN SHOWER LIGHTS SWINGING FROM A BEAM OVER A LONG DINING TABLE. THIS IS ONE OF THE MOST EFFECTIVE METHODS WE HAVE FOUND OF MANAGING THE LIGHTS IN A DINING ROOM.

A CONVENIENT AND WELL-EQUIPPED KITCHEN THAT SIMPLIFIES THE HOUSEWORK

EACH room in the house has its distinct and separate function in the domestic economy. Therefore it should be remembered that before any room can attain its own distinctive individuality everything put into it must be there for some reason and must serve a definite purpose in the life that is to be lived and the work that is to be done in that room. Take for example the kitchen, where the food for the household must be prepared and where a large part of the work of the house must be done. This is the room where the housewife or the servant maid must be for the greater part of her

good fortune to associate such a room with their earliest recollections of home. No child ever lived who could resist the attraction of such a room, for a child has, in all its purity, the primitive instinct for living that ruled the simpler and more wholesome customs of other days. In these times of more elaborate surroundings the home life of the family is hidden behind a screen and the tendency is to belittle that part of the household work by regarding it as a necessary evil. Even in a small house the tendency too often is to make the kitchen the dump heap of the whole household, a place in which to do what cooking and

CORNER OF THE KITCHEN SHOWING BUILT-IN CUPBOARD AND SINK.

time day after day, and the very first requisites are that it should be large enough for comfort, well ventilated and full of sunshine, and that the equipment for the work that is to be done should be ample, of good quality and, above all, intelligently selected. We all know the pleasure of working with good tools and in congenial surroundings; no more things than are necessary should be tolerated in the kitchen and no fewer should be required.

We cannot imagine a more homelike room than the old New England kitchen, the special realm of the housewife and the living room of the whole family. Its spotless cleanliness and homely cheer are remembered as long as life lasts by men and women who have had the

dishwashing must be done and to get out of as soon as possible. In such a house there is invariably a small, cheap and often stuffy dining room, as cramped and comfortless as the kitchen and yet regarded as an absolute necessity in the household economy. Such an arrangement is the result of sacrificing the old-time comfort for a false idea of elegance and its natural consequence is the loss of both.

In the farmhouse and the cottage of the workingman, where the domestic machinery is comparatively simple, cheerful and homelike, the kitchen,—which is also the dining room of the family and one of its pleasantest gathering places,—should be restored to all its old-time comfort and convenience. In

A CONVENIENT AND WELL-EQUIPPED KITCHEN

planning such a house it should come in for the first thought instead of the last and its use as a dining room as well as a kitchen should be carefully considered. The hooded range should be so devised that all odors of cooking are carried off and the arrangement and ventilation should be such that this is one of the best aired and sunniest of all the rooms in the house.

Where social relations and the demands of a more complex life make it impossible for the house mistress to do her own work and the kitchen is necessarily more separated from the rest of the household, it may easily be planned to meet the requirements of the case without losing any of its comfort, convenience, or suitability for the work that is to be done in it. Modern science has made the task very easy by the provision of electric lights, open plumbing, laundry conveniences, and hot and cold running water, so that the luxuries of the properly arranged modern kitchen would have been almost unbelievable a generation ago. Even if the kitchen is for the servant only, it should be a place in which she may take some personal pride. It is hardly going too far to say that the solution of the problem of the properly arranged kitchen would come near to being the solution also of the domestic problem.

The properly planned kitchen should be as open as possible to prevent the accumulation of dirt. Without the customary "glory holes" that sink and other closets often become, gen-

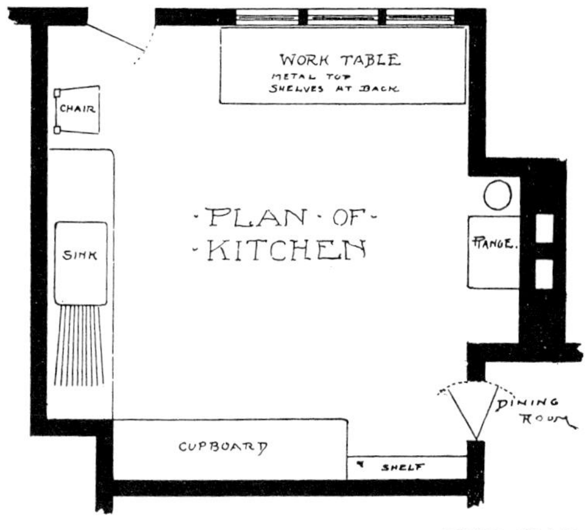

FLOOR PLAN.

uine cleanliness is much easier to preserve and the appearance of outside order is not at all lessened. In no part of the house does the good old saying, "a place for everything and everything in its place," apply with more force than in the kitchen. Ample cupboard space for all china should be provided near the sink to do away with unnecessary handling and the same cupboard, which should be an actual structural feature of the kitchen, should contain drawers for table linen, cutlery and smaller utensils, as well as a broad shelf which provides a convenient place for serving. The floor should be of cement and the same material may be used in tiled pattern for a high wainscot, giving a cleanly and pleasant effect.

Published in The Craftsman, September, 1905.

RANGE SET IN A RECESS TO BE OUT OF THE WAY AND WORK TABLE PLACED JUST BELOW A GROUP OF WINDOWS.

THE TREATMENT OF WALL SPACES SO THAT A ROOM IS IN ITSELF COMPLETE AND SATISFYING

SO much of the success of any scheme of interior decoration or furnishing depends upon the right treatment of the wall spaces that we deem it best to take up this subject more in detail than it has been possible to do in the general descriptions of the houses or even of the separate rooms.

It goes without saying that we like the friendly presence of much wood and are very sensible of the charm of beams, wainscots and built-in furnishings which are a part of the house itself and so serve to link it closer to the needs of daily life. Bare wall spaces, or those covered with pictures and draperies which are put there merely for the purpose of covering them, are very hard to live with. But wall spaces that provide bookcases, cupboards, built-in seats for windows, fireside and other nooks are used in a way that not only gives to them the kind of beauty and interest which is theirs by right, but makes them of practical value in the life of the household, as such furnishings mean great convenience, economy of space and the doing away with many pieces of furniture which might otherwise be really needed, but which might give the appearance of crowding that is so disturbing to the restfulness of a room.

When the walls are rightly treated, it is amazing how little furniture and how few ornaments and pictures are required to make a room seem comfortable and homelike. The treatment of wall spaces in itself may seem but a detail, yet it is the keynote not only of the whole character of the house but of the people who live in it. We hear much criticism of the changing and remodeling which is deemed necessary every year or two because a house must be "brought up to date" or because the owners "grow so tired of seeing one thing all the time." Yet both of these reasons are absolutely valid so far as they go, for the

Published in The Craftsman, October, 1904.

A HIGH WAINSCOT MADE WITH RECESSES TO HOLD CHOICE BITS OF METAL OR EARTHENWARE. THIS IS ESPECIALLY BEAUTIFUL IF CARRIED OUT IN CHESTNUT OR GUMWOOD TREATED IN THE CRAFTSMAN MANNER.

majority of houses are in themselves so uninteresting that it is little wonder that the people who live in them have always a sense of restlessness and discontent, and that they are always doing something different in the hope that eventually they may find the thing which satisfies them.

We believe that the time to put thought into the decoration of a home is when we first begin to draw up the plans, and that the first consideration in each room should be the adjustment of the wall spaces so that there is not a foot of barren or ill-proportioned space in the entire room. It is true that utility and the limitations of the plan are necessarily the first considerations; that the ceilings of all the rooms on one story must be of uniform height in a house where the expense of construction is a thing to be considered; that windows must be placed where they will admit the most light and that doors are meant to serve as means of communication between rooms or with the outer world. Yet working strictly within these limitations, it is quite possible to adjust the height of each room so

Published in The Craftsman, June, 1905.

WALL DIVIDED INTO PANELS BY STRIPS OF WOOD.

that, no matter what may be its floor space, to all appearances its proportions are entirely harmonious; to place doors and windows so that, instead of being mere holes in the wall, they become a part of the whole structural scheme, and to see that in shape and proportions as well as in position they come into entire harmony with the rest of the room.

Published in The Craftsman, October, 1907.

LOW WAINSCOT WITH BROAD PANELS. NOTE THE PLACING OF THE WINDOW SO THAT IT REALLY FORMS A DECORATIVE PANEL IN THE WALL SPACE ABOVE.

Published in The Craftsman, June, 1905.

ATTRACTIVE TREATMENT OF WALLS IN A BEDROOM OR WOMAN'S SITTING ROOM.

Naturally, in considering the treatment of the wall spaces, the most important feature is the woodwork, especially if the room is to be wainscoted. Where this is possible, we would always recommend it, particularly for the living rooms of a house, as no other treatment of the walls gives such a sense of friendliness, mellowness and permanence as does a generous quantity of woodwork. The larger illustrations reproduced here give some idea of what we mean and of what may be done with wall spaces when it is possible to use much wood in the shape of wainscot and beams. It will be noted that in each case the wall is of the same height; yet owing to the treatment of the spaces, each one appears to be different. Also note the way in which windows, doors and fireplace form an integral part of the structural scheme and how they are balanced by the wall spaces around them so that the whole effect is rather that of a well planned scheme of structural decoration than of the introduction of a purely utilitarian feature.

When we speak of the friendliness of woodwork, however, we mean woodwork that is so finished that the friendly quality is apparent, —which is never the case when it is painted or stained in some solid color that is foreign to the wood itself, or is given a smooth glassy polish that reflects the light. When this is done the peculiar quality of woodiness, upon which all the charm of interior woodwork depends, is entirely destroyed and any other material might as well be used in the place of it. In a later chapter we purpose to deal more

Published in The Craftsman, October, 1907.

TREATMENT OF WAINSCOTED WALL IN A LIVING ROOM WHERE THE PANELING IS REPEATED IN THE FRIEZE AND THE FIREPLACE IS PERFECTLY PROPORTIONED IN RELATION TO THE WALL SPACES ON EITHER SIDE.

fully with the question of finishing interior woodwork so that all its natural qualities of color, texture and grain are brought out by a process which ripens and mellows the wood as if by age without changing its character at all. Here it is sufficient to say that any of our native woods that have open texture, strong grain and decided figure,—such as oak, chestnut, cypress, ash, elm or the redwood so much used on the Pacific coast,—are entirely suitable for the woodwork of rooms in general use, and that each one of them may be so finished that its inherent color quality is brought out and its surface made pleasantly smooth without sacrificing the woody quality that comes from frankly revealing its natural texture.

The first illustration (page 144) shows a wainscot that is peculiarly Craftsman in design. The panels are very broad and what would be the stiles in ordinary paneling are even broader. At the top of each panel is a niche in which may be set some choice bit of pottery or metal work that is shown to the best advantage by the wood behind it and that serves to give the accents or high lights to the whole color scheme of the room. The

Published in The Craftsman, June, 1905.

TREATMENT OF PLAIN WALLS WITH LANDSCAPE FRIEZE.

wall space above is of plain sand-finished plaster that may either be left in the natural gray or treated with a coat of shellac or wax which carries the color desired. The rough texture of the plaster has the effect of seeming to radiate color, while it absorbs the light instead of reflecting it as from a smoothly polished surface, and when the color is put on lightly enough to be a trifle uneven instead of a dead solid hue without variation of any sort, there is a chance for the sparkle and play of light which at once adds life and interest.

Published in The Craftsman, October, 1907.

WALL WITH A HIGH WAINSCOT IN WHICH THE DOOR AND WINDOW ARE MADE A PART OF THE STRUCTURAL DECORATION: THE LEADED PANELS IN WINDOW AND DOOR ADD MUCH TO THE BEAUTY OF THE ROOM.

Published in The Craftsman, August, 1905.

TREATMENT OF WALLS IN A NURSERY. PLAIN ROUGH PLASTER BELOW AND A SHADOWY SUGGESTION OF A FOREST IN THE FRIEZE. BY THIS ARRANGEMENT THE SURFACE OF THE LOWER WALL IS EASILY KEPT CLEAN AND YET ALL APPEARANCE OF BARRENNESS IS AVOIDED.

Published in The Craftsman, August, 1905.

ANOTHER SUGGESTION FOR THE TREATMENT OF NURSERY WALLS, SHOWING A PICTURE DADO ILLUSTRATING NURSERY TALES AND A BLACKBOARD BUILT INTO THE WAINSCOT WITHIN EASY REACH OF THE LITTLE ONES.

FLOORS THAT COMPLETE THE DECORATIVE SCHEME OF A ROOM

ONE of the most important elements in the success of a room designed to be beautiful as a whole in structure and color scheme, is the floor. Whether it be a more or less elaborate parquet floor or one made simply of plain boards, it must be in harmony with the color chosen for the wood trim of the room. Also it should invariably be at least as dark as the woodwork, if the effect of restfulness is to be preserved. A floor that strikes a higher note of color than the woodwork above it, even if it be otherwise harmonious in tone, gives the room a top-heavy, glaring effect that no furniture or decoration will remove.

Full directions for finishing floors will be given later in the chapter on wood finishes. While the Craftsman method of finishing woodwork differs widely from others, it does not apply so much to the floor, for here a filler should be used for precisely the same reason that it should be avoided in the treatment of furniture and woodwork, as it destroys the texture of the wood by covering it with a glassy, smooth and impervious surface. Texture is not needed in the wood of a floor, which should be entirely smooth and non-absorbent.

The first of the three floors illustrated here is meant to complete the color scheme of a room in which the woodwork is of silver-gray maple and the furniture and decorations are in delicate tones such as would naturally harmonize with gray. The floor is very simple in design, having a plain center of silver-gray maple that is finished exactly like the woodwork of the room. Around the edge is a wide border of "mahajua," a beautiful Cuban hardwood, close and smooth in grain and left in its natural color, which is a greenish gray slightly darker than the finish of the maple.

The second floor is made of quartered oak in the natural color, and the boards are bound together with keys of vulcanized oak. Where the floor is stained to match the woodwork in tone, the color value of boards and keys will remain the same, as the vulcanized oak keys will simply show a darker shade of whatever color is given the boards of plain oak. The last illustration shows a floor of quartered oak in the natural color combined with vulcanized oak and white maple to form a border in which a primitive Indian design appears.

Published in The Craftsman, October, 1905.

A FLOOR OF SILVER-GRAY MAPLE AND MAHAJUA.

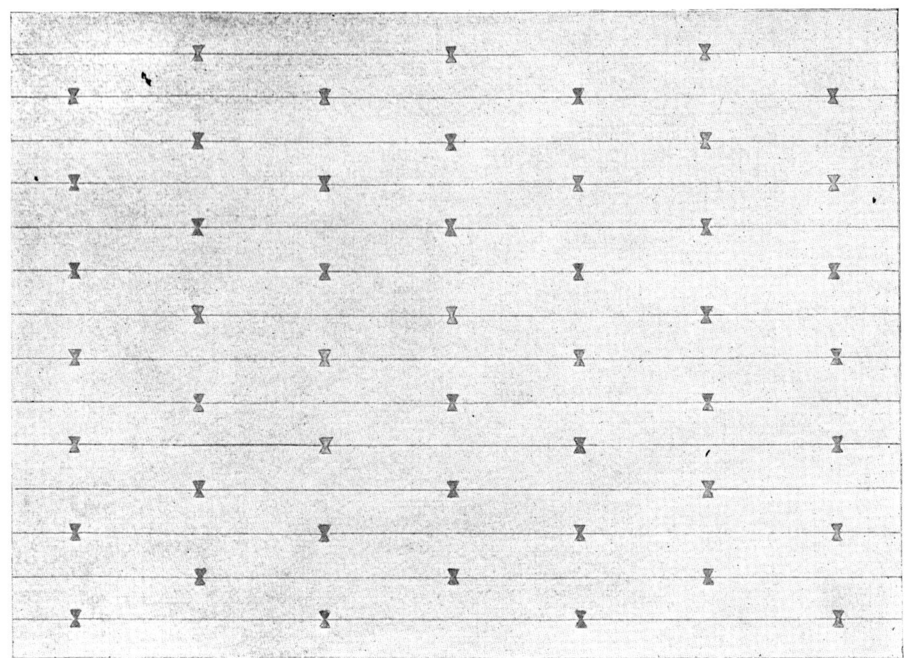

Published in The Craftsman, October, 1905.

FLOOR OF NATURAL OAK INLAID WITH KEYS OF VULCANIZED OAK.

Published in The Craftsman, October, 1905.

FLOOR OF OAK INLAID WITH MAPLE. BORDER IN INDIAN DESIGN.

AN OUTLINE OF FURNITURE-MAKING IN THIS COUN-TRY: SHOWING THE PLACE OF CRAFTSMAN FURNI-TURE IN THE EVOLUTION OF AN AMERICAN STYLE

THIS book is meant to give a comprehensive idea of the elements that go to make up the typical Craftsman home. Therefore at least one chapter must be devoted to Craftsman furniture, for in the making of this we first gave form to the idea of home building and furnishing which we have endeavored to set forth. For this reason, and because the furniture has so far remained the clearest concrete expression of the Craftsman idea, we are here illustrating a few of the most characteristic pieces that

scope of their experience, for, after the first primitive days of the Pilgrim Fathers in New England and the earliest settlers in the South, the life of the Colonists was modeled closely upon that of the old country and this life naturally found expression in their dwellings and household belongings. Therefore the Colonial style was so close to the prevailing style of the eighteenth century that it may be regarded as practically the same thing.

After the end of the Colonial period, and during the swift expansion that followed the

ONE OF THE LARGEST AND MOST MASSIVE OF THE CRAFTSMAN SETTLES; MADE OF FUMED OAK; SOFT LEATHER SEAT.

serve to show all the essential qualities of the style.

In order that the reader may understand clearly the reasons which led to the making of Craftsman furniture, and its place in the evolution of a distinctively American style that bids fair eventually to govern the great majority of our dwellings and household belongings, we will first briefly review the history of furniture making in this country. With the older styles, such as the English and the Dutch Colonial, we have little to do. They were importations from older civilizations, as were the Colonists themselves, and they expressed the life of the mother country rather than that of the new. When we first began to make furniture in this country, the cabinet-makers naturally followed their old traditions and made the kind of furniture which most appealed to them and which came within the

Revolution, there was inevitably a return to the primitive. Importations from the old world were no longer popular and while the houses of the wealthy were still furnished with the graceful spindle-legged mahogany pieces of earlier days, most of the people were forced to content themselves with much plainer and more substantial belongings. Little chair factories sprang up here and there, especially in Maine, Vermont and Massachusetts, and these supplied the great demand for the plain wooden chairs that we now call kitchen chairs, and the cane-seated chairs which were usually reserved for use in the best room. As the demand increased with the increasing population, the alert and resourceful New Englander began to invent machinery which would increase his output. As a consequence, the business of chair making made rapid growth, but the primitive

FURNITURE MAKING IN THIS COUNTRY

AN ARM-CHAIR AND ROCKER THAT ARE BUILT FOR SOLID COMFORT AS WELL AS DURABILITY.

beauty of the hand-made pieces was lost. The Windsor chairs, with their perfect proportions, subtle modeling and slender legs shaped with the turning lathe, became a thing of the past, for in the factories it was necessary from a business point of view to effect the utmost savings in material and also to consider the limitations of the machinery of that day. The object of the manufacturer naturally was to turn out the greatest possible quantity of goods with the least possible amount of labor and expense, and the result was so many modifications of the original form that the factory-made chairs soon become commonplace. When machines were invented to take the place of hand turning and carving, it was inevitable that vulgarity should be added to the commonplaceness, be-

LARGE CRAFTSMAN LOUNGING CHAIR.

cause it is so easy to disguise bad lines with cheap ornamentation.

Side by side with these chair factories another furniture industry was springing up, mainly in the Middle West because that was the black walnut country and black walnut was the material most in demand for the more elaborate furniture. At the same time that the New Englander was evolving from the artisan who carried on his work with the aid of a little water mill, to a manufacturer who owned a chair factory run by machinery, a number of German cabinetmakers who had settled in Indiana and the neighboring states were accumulating, by means of industry and thrift, enough means to set up general

LARGE OCTAGONAL TABLE, TOP COVERED WITH HARD LEATHER. DESIGNED FOR LIBRARY OR LIVING ROOM.

furniture factories, which supplied the country with black walnut "parlor suits," upholstered with haircloth, repps or plush, while the New Englander remained content to furnish it with dining room and kitchen chairs.

This period in our furniture corresponds with the architectural phase in this country which has aptly been termed the "reign of terror," but we are in some measure consoled for the hideous bad taste of it all by the reflection that it was contemporary with the early and mid-Victorian period in England, a term that everywhere stands for all that is ugly, artificial and commonplace in household art. It was succeeded by the first of the Grand Rapids furniture, which was in some measure a change for the better. Tempted by the success of the German furniture makers, the

FURNITURE MAKING IN THIS COUNTRY

shrewd New England manufacturers, with their superior knowledge of machinery, managed to plant themselves in t h e Middle West and to distance their competitors. The center of these new manufacturing interests was then in Grand Rapids, Michigan, so that the new style of furniture which was produced came to be known as G r a n d Rapids furniture. It was plainer than the black walnut furniture and was fashioned more after the Colonial models, but the best features were speedily lost in the ornamentation with which it was overlaid, as well as in the modification and adaptation of the earlier forms by a new generation of designers, who had studied foreign furniture and so gained a smattering of the traditional styles which they proceeded to apply to the creation of "novelties." About this time the large department stores sprang up and, as they very soon became the principal retailers, they naturally assumed control of the furniture that was made. The demand for novelties was unceasing and the designer was at the beck and call of the traveling salesman, who in his turn was compelled to supply a ceaseless stream of new attractions to the head of the furniture department,—whose business it was constantly to whet the public

LOW ROCKER AND DROP-LEAF SEWING TABLE WITH THREE DRAWERS; THE UPPER ONE HAVING A SLIDING TRAY MADE OF CEDAR WITH COMPARTMENTS FOR SPOOLS.

appetite for further novelties.

The greater part of the demand thus created was satisfied by the Grand Rapids f u r n i t u r e, but as wealth and culture increased, and people became more and more familiar with European homes and European luxuries. the new vogue for the "period" furniture sprang up among the richer c l a s s, and some of the factories turned their attention to endeavoring to duplicate the several styles of French and English furniture of the seventeenth and eighteenth centuries. These factories are still running, some of them being employed in turning out the closest imitation they can make of the "period" furniture and others in reproducing Colonial models.

While we were doing these things in America, Ruskin and Morris had been endeavoring to establish in England a return to handicrafts as a means of individual expression along the several lines of the fine and industrial arts. This gave rise over there to the Arts and Crafts movement, which was based chiefly upon the expression of untrammeled individualism. Much furniture was made,—some of it good, but a great deal of it showing the eccentricities of personal fancy un-

ROUND TABLE THAT IS WELL ADAPTED TO GENERAL USE.

CHESS OR CHECKER TABLE HAVING TOP COVERED WITH HARD LEATHER MARKED OFF INTO SQUARES FOR THE BOARD.

A BOOKCASE THAT IS A GOOD EXAMPLE OF THE DECORATIVE USE OF PURELY STRUCTURAL FEATURES.

permanent style in English furniture for the reason that they have striven for a definite and intentional expression of art that was largely for art's sake and had little to do with satisfying the plain needs of the people. Although the founders of the movement held and preached the doctrine that all vital art necessarily springs from the life of the people, it is nevertheless recognized even by their followers that in practice such expression as they advocate belongs to the artist alone and that the people care very little about it.

A LIGHT WRITING TABLE FOR A LIVING ROOM OR SMALL SITTING ROOM.

modified by any settled standards. It was a move in the right direction because it meant a return to healthy individual effort and a revolt from the dead level established by the machines. But the Arts and Crafts workers have not succeeded in establishing another

It was during this same period that the movement called *L'Art Nouveau* sprang up in France and for a time attained quite a vogue under the leadership of Bing. Belgium followed suit with a rather heavier and more pronounced interpretation of the distinctive features of this style and for a few years the plant forms and swirling lines that distinguished *L'Art Nouveau* productions were very popular. In Germany and Austria the art students and others of the more restless spirits who were constantly in revolt from the established styles determined to outdo the French and accordingly established government schools for the teaching of a definite style, which was called New Art or Secessionist and which

RUSH SEATED CHAIR AND DROP LEAF TABLE WITH SEPARATE WRITING CABINET.

contained some of the features of *L'Art Nouveau* and others that were borrowed from the English Arts and Crafts and also from ancient Egyptian forms of art. The French school has already failed in the efforts to establish a permanent style, and the indications are that the efforts of the German and Austrian Secessionists will prove equally futile, because in both cases the workers have merely attempted to do something different; to evolve a new thing by combining the features of the old. In other words, they began at the top instead of beginning at the bottom and allowing the style to develop naturally

A LARGE WRITING DESK FOR THE LIBRARY OR WORKROOM.

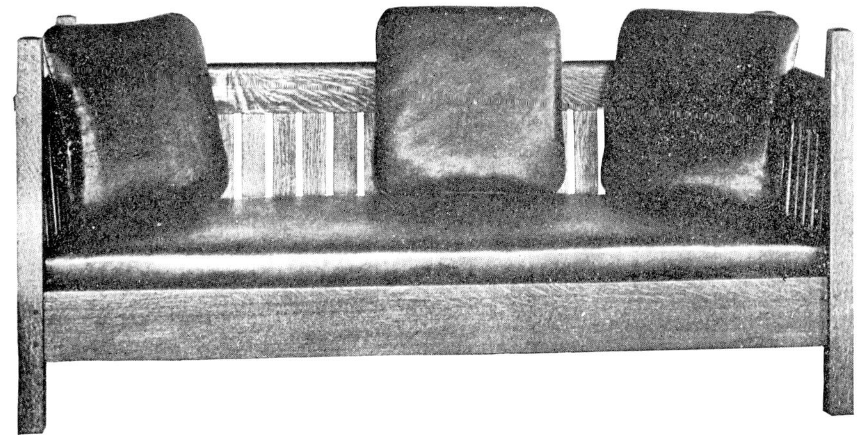

A TYPICAL CRAFTSMAN LOUNGING CHAIR.

from the sure foundation of real utility. The leaders succeeded in making things that, whatever their relative merits, were a new departure; but this once made, it stood as a completed achievement that might be imitated, but could hardly be developed, as it lacked the beginnings of healthy growth.

But during the same period in this country things were on a different basis. Out of the chaos of ideals and standards which had naturally resulted from the rapid growth of the young nation, a vigorous and coherent national spirit was being developed, and amid the general turmoil and restlessness attendant upon swift progress and expansion, it became apparent that we were evolving a type of people distinct from

A LARGE OAKEN SETTLE UPHOLSTERED WITH CRAFTSMAN SOFT LEATHER. THE PILLOWS ARE COVERED WITH THE STILL SOFTER AND MORE FLEXIBLE SHEEPSKIN.

FURNITURE MAKING IN THIS COUNTRY

all others,—a type essentially American. And the distinguishing characteristic of this type is the power to assimilate so swiftly the kind of culture which leads to the making of permanent standards of life and art that it is hardly to be compared with what might seem to be the corresponding class in other countries. Such Americans have fundamental intelligence and the power of discrimination, and the direct thinking that results from these qualities inevitably produces a certain openness of mind that responds very quickly to

ness alone. In this country, where we have no monarchs and no aristocracy, the life of the plain people is the life of the nation; therefore, the art of the age must necessarily be the art of the people. Our phases of imitation and of vulgar desire for show are only a part of the crudity of youth. We have not yet outgrown them and will not for many years; but as we grow older and begin to stand on our own feet and to cherish our own standards of life and of work and therefore of art, we show an unmistakable tendency to

A GROUP OF CRAFTSMAN SPINDLE FURNITURE. THIS IS QUITE AS STRONG AND DURABLE AS THE MORE MASSIVE PIECES BUT IS A LITTLE LIGHTER IN APPEARANCE.

anything which seems to have a real and permanent value.

This quality was shown in the immediate recognition and welcome accorded to Craftsman furniture when we first introduced it ten years ago. Like the Arts and Crafts furniture in England, it represented a revolt from the machine-made thing. But there was this difference: The Arts and Crafts furniture was primarily intended to be an expression of individuality, and the Craftsman furniture was founded on a return to the sturdy and primitive forms that were meant for useful-

get away from shams and to demand the real thing.

And to an American the real thing is something that he needs and understands. The showroom quality is all very well when it comes to proving how much money he has or to establishing a reputation for owning things that are just as good as his neighbor's. But for use he wants the things that belong to him,—the things that are comfortable to live with; that represent a good investment of his money and have no nonsense about them. Furthermore the true American likes to know

192

how things are done. His interest and sympathy are immediately aroused when he sees something that he really likes and knows to be a good thing, if he is able to feel that, if he wanted it and had the time, he could make one like it himself.

So strong is this national characteristic that it is hardly overstating the case to say that in America any style in architecture or furniture would have to possess the essential qualities of simplicity, durability, comfort and convenience and to be made in such a way that

SMALL WRITING DESK FOR A WOMAN'S SITTING ROOM OR FOR THE LIVING ROOM

the details of its construction can be readily grasped, before it could hope to become permanent. We are not so many generations removed from our pioneer forefathers that we have grown entirely out of their way of getting at things. We may not always stop to think about them, but when we do our thought is apt to be fairly sound and direct. The prevalence of cheap, showy, machine-made things in our houses is due chiefly to the lack of thought that

A BIG DEEP CHAIR THAT MEANS COMFORT TO A TIRED MAN WHEN HE COMES HOME AFTER THE DAY'S WORK.

takes on trust the word of the dealers, and every year brings us more abundant proof that they do not in any way represent the real tastes and standards of the people. All machine-made imitations of furniture which belonged to another country and another age and represented the life of a totally different people, are alike to the average American. If he can get them cheap, he has at least the satisfaction of feeling that they make a pretty good otuward show for the money; if they are expensive, there is something in being able to afford them and to know that his house has in it rooms which are fairly

NO BETTER EXAMPLES OF THE CRAFTSMAN STYLE CAN BE FOUND THAN ARE SHOWN IN THIS CHAIR AND ROCKER.

FURNITURE MAKING IN THIS COUNTRY

CRAFTSMAN SIDEBOARD WHERE WROUGHT IRON PULLS AND HINGES ARE USED IN A DECORATIVE WAY.

successful imitations of the rooms in French or English palaces two or three hundred years ago. But in all this there is no real thought and nothing that approaches it. It is only when a thing has the honest primitive quality that reveals just what it is, how it is made and what it is made for, that it comes home to us as something which possesses an individuality of its own. It is not an elaborate finished thing made by machinery with intricate processes which we cannot understand and about which we do not care in the least; it is something that we might make with our own hands. Therefore it is something that sets us to thinking and establishes a point of contact from which springs the essentially human qualities of interest and affection. Understanding just how it is made, we are in a position to appreciate exactly what the artisan has done and how well he has done it. From this understanding comes the personal interest in good work that alone gives the vital quality which we know as art.

Many people misunderstand the meaning of the word primitiveness, mistaking it for crudeness, but the word is used here to express the directness of a thing that is radical instead of derived. In our understanding of the term, the primitive form of construction is that which would naturally suggest itself to a workman as embodying the main essentials of a piece of furniture, of which the first is the straightforward provision for practical need. Also we hold that the structural idea should be made prominent because lines which clearly define their purpose appeal to the mind with the same force as does a clear concise statement of fact. This principle is the basis from which the Craftsman style of furniture has been developed. In the beginning there was no thought of creating a new style, only a recognition of the fact that we should have in our homes something better suited to our needs and

CRAFTSMAN DINING TABLE AND TWO OF THE MORE MASSIVE DINING CHAIRS, ONE UPHOLSTERED IN HARD LEATHER STUDDED WITH DULL BRASS NAILS AND THE OTHER MADE OF PLAIN OAK WITH A HARD LEATHER SEAT.

FURNITURE MAKING IN THIS COUNTRY

more expressive of our character as a people than imitations of the traditional styles, and a conviction that the best way to get something better was to go directly back to plain principles of construction and apply them to the making of simple, strong, comfortable furniture that would meet adequately everything that could be required of it.

Because Craftsman furniture expresses so clearly the fundamental sturdiness and directness of the true American point of view, it follows that in no other country and under no other conditions could it have been produced at the present day. The history of art shows us that a new form of expression never develops from the top and that nothing permanent is ever built upon tradition. When a style is found to be original and vital it is a certainty that it has sprung from the needs of the plain people and that it is based upon the simplest and most direct principles of construction. This is always the beginning and a style that has in it sufficient vitality to endure, will grow naturally as one worker after another feels that he has something further to express. In making Craftsman furniture we went back to the beginning,

seeking the inspiration of the same law of direct answer to need that animated the craftsmen of an earlier day, for it was suggested by the primitive human necessity of the common folk. It is absolutely plain and unornamented, the severity of the style marking a point of departure from which we believe that a rational development of the decorative idea will ultimately take place.

WILLOW CHAIRS AND SETTLES WHICH HARMONIZE WITH THE MORE SEVERE AND MASSIVE FURNITURE MADE OF OAK

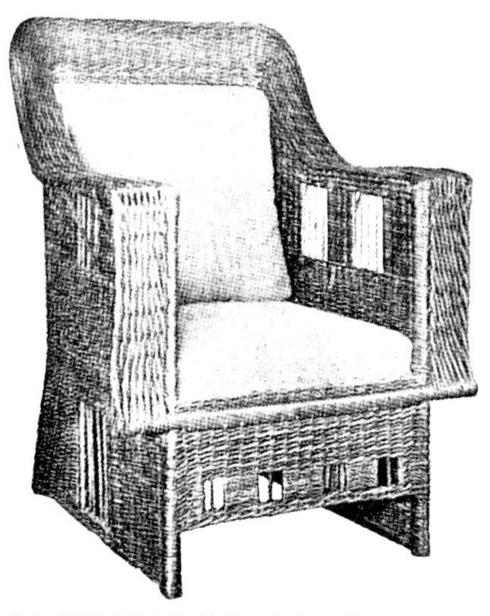

AN ARM CHAIR OF WOVEN WILLOW.

THE opinion is frequently expressed with regard to Craftsman furniture that it is all very well for the library, den or dining room, but that an entire house furnished with it would be apt to appear too severe and monotonous in its general effect. While naturally we feel that Craftsman furniture is equally suitable for every room in the house, we are aware that there is precisely the same element of truth in this criticism that it holds when applied to any kind of furniture. The point is that too much of any one thing is apt to be monotonous, and the way we avoid that fault in a Craftsman house is to make the furniture entirely a secondary thing and keep it as little obtrusive as possible, so that each piece sinks into its place in the picture and becomes merely a part of the general impression, instead of standing out as a separate article.

In the Craftsman houses we do away with a great deal of the movable furniture by the use in its place of built-in fittings, which are made a part of the structure of the house. As these include window seats, fireside seats, settles, bookcases, desks, sideboards, china cupboards and many other things, it will easily be seen that their presence not only adds to the structural interest and beauty of the room itself, but makes it possible to dispense with much of the furniture which would otherwise be needed. For the rest, we use Craftsman furniture where it is necessary to have pieces of wood construction, but we relieve any possible severity of effect by a liberal use of willow settles and chairs which afford the best possible foil to the austere lines, massive forms and sober coloring of the oak. We select willow for this use rather than rattan, because, while all such furniture is necessarily handmade, the rattan pieces are usually patterned after the elaborate effects that we have learned to associate with machine-made goods, and so have none of the natural interest that is a part of something which grows under the hand and is shaped as simply as possible to meet the purpose for which it is intended.

The charm of willow is that it is purely a handicraft, and obviously so. A rattan chair or settle may be twisted into any fan-

A HIGH-BACK SETTLE OF WILLOW THAT HARMONIZES ADMIRABLY WITH THE GENERAL CHARACTER OF CRAFTSMAN FURNITURE.

tastic form, but willow furniture is essentially of basket construction. Our idea in making the kind of willow furniture illustrated here was to gain something based upon the same principles of construction that characterize our oak furniture; that is, to secure a form that should suggest the simplest basket work and the flexibility of lithe willow branches and yet be as durable as any of the heavy oak furniture which is emphatically of wood construction.

Consequently these pieces are basketry pure and simple and have an elastic spring under the pressure of the body that suggests the flexibility of baskets such as are woven by the fireside or on the back porch at the edge of the garden. The making of willow furniture as a handicraft is rather a hobby with us, for willow is a material beloved of the craftsman and the work is very interesting and comparatively easy to do. The trouble is that so many people are inclined to overdo it and to make out of woven willow the kind of furniture that demands wood construction. Seat furniture alone is permissible in willow and yet we frequently see tables, racks and stands of various kinds, and even the front of a bureau or a dresser, made of this material. Such misuse is a pity, the more that it tends to create a prejudice again against willow furniture as a whole.

The pieces shown here hold in their beauty of form and color evidences of the personal interest of the worker. The willow has been so finished that the surface has the sparkle seen in the thin branches of the growing tree as it becomes lustrous with the first stirring of the sap. This natural sparkle on the surface of willow has all

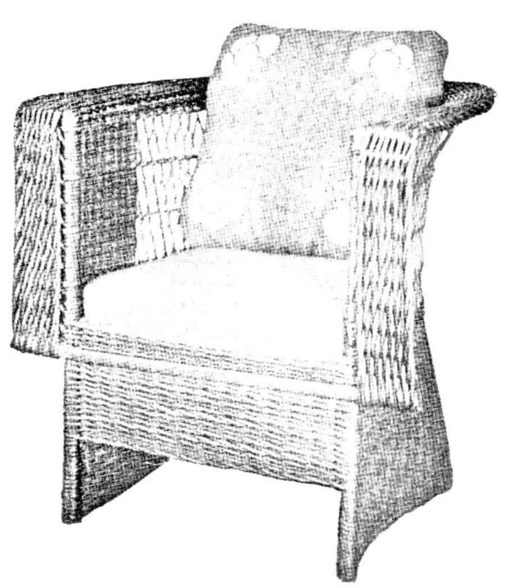

WILLOW CHAIR MADE ON A LOWER AND BROADER MODEL.

the intangible silvery shimmer of water in moonlight. This is lost absolutely when the furniture made of it is covered with the usual opaque enamel, which not only hides the luster of the surface but gives the effect of a stiff uncompromising construction in which the pliableness of the basket weave is entirely obliterated and all the possible interesting variations of tone are lost under the smooth surface.

We finish our willow furniture in two colors; one gives the general impression of green, but it is really a variation of soft wood tones, brown and green, light and dark, as the texture of the withes has been smooth or rough. In this way the silvery luster of the willow is left undisturbed and the color beneath is like that of fresh young bark. The other color is golden brown in which there is also a suggestion of spring-like gray and green.

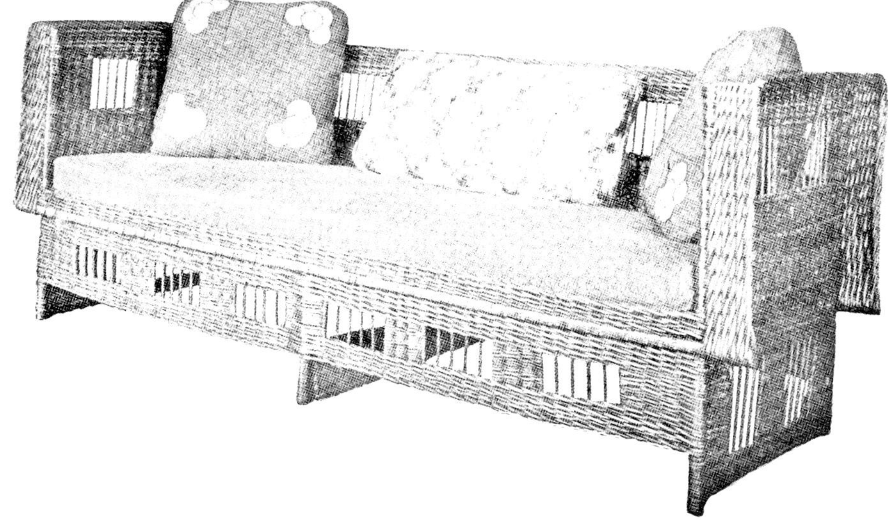

VERY LARGE WILLOW SETTLE MADE AFTER A DESIGN THAT WE HAVE FOUND MOST SATISFACTORY IN RELATION TO THE REGULAR CRAFTSMAN FURNITURE OF OAK.

CRAFTSMAN METAL WORK: DESIGNED AND MADE ACCORDING TO THE SAME PRINCIPLES THAT RULE THE FURNITURE

IN a room decorated according to Craftsman ideas,—especially if it be furnished with Craftsman furniture,—it is of the utmost importance that the metal accessories should be of a character that fits into the picture. We found out very soon after we began to make the plain oak furniture that even the best of the usual machine-made and highly polished metal trim was absurdly out of place, and that in order to get the right thing it was necessary to establish a metal-work department in the Craftsman Workshops where articles of wrought metal in plain rugged designs and possessing the same structural and simple quality as the furniture could be made. We began with such simple and necessary things as drawer and door pulls, hinges and escutcheons, but with a work so interesting and so full of possibilities as this one thing inevitably leads to another, and our metal workers were soon making in hand-wrought iron, copper and brass all kinds of household fittings, such as lighting fixtures, fire sets, and other articles that were decorative as well as

COPPER-FRAMED LANTERN THAT IS INTENDED TO HANG FROM A BRACKET ATTACHED TO THE WALL.

LARGE LANTERN THAT IS BEST FITTED FOR USE IN AN ENTRANCE HALL OR VERANDA.

useful, and that showed the same essential qualities as the furniture.

Since then we have not only made all manner of metal furnishings ourselves, but through the pages of THE CRAFTSMAN we have warmly encouraged amateur workers to do the same thing and have given for their use a number of models as well as full directions regarding methods of working and the necessary equipment for doing all kinds of simple metal work at home. Under the inspiration of these suggestions and directions, a number of readers of THE CRAFTSMAN have set up little home workshops and have succeeded in making many pieces that show originality and merit. In fact, metal work is one of the most interesting of the crafts to the home worker who possesses skill and taste and, above all, a genuine interest in making for himself the things that are needed either for use or ornament at home, and anyone who takes it up and discovers its possibilities is likely to go on with it indefinitely. Instruction in the technicalities is easily obtained from any blacksmith who can teach the rudiments of handling iron, or from any working jeweler or coppersmith who is able to give the necessary personal supervision to the first efforts of a worker in brass or copper. Given even a little ingenuity and handiness with tools, it might be possible to dispense even with this instruction and to work out each problem as it comes up; learning by doing, in the simple way of the handicraftsman of old.

It ought to be possible for such home workers to make everything necessary for the fireplace, including shovels and tongs, andirons, fenders, coal buckets and even fireplace hoods, although the last named might be a fairly ambitious undertaking for an amateur. One needs but little imagination to realize the interest and charm that would attach to a comfortable fireside nook that had been furnished in this way, and the same principle applies to every one of the smaller articles of furniture in the home. For example, it is not at all hard to make from either brass or copper a tray or an umbrella stand, a simple vase or metal jug or a jardiniére, and the quality of really wonderful; that worker takes fine himself

SMALL SQUARE LANTERN MEANT TO HANG FROM THE CEILING OR AN OVERHEAD BEAM.

SQUARE LANTERN WITH AN UNUSUALLY DECORATIVE COPPER FRAME.

possible the need for which the just as well as he can, keeping mon to metal workers of artifi- wrought" effect by putting ham- business to be, leaving the exaggerating into crudity the if rightly used, give to a terest and charm. Much of way the metal is finished. wrought-iron work is

good designs that meet as directly as article is made, and then makes it free from the temptation so com- cially heightening the "hand- mer dents where they have no edges rough and generally traces of workmanship which, piece such a human in- the effect depends upon the For example, all of our finished in a way that has

ELECTROLIER IN FUMED OAK AND HAMMERED COPPER, ESPECIALLY DESIGNED FOR HANGING RATHER LOW OVER A DINING TABLE.

long been known in England as "armor bright." This is a very old process used by the English armorers, whence it derives its name, and its peculiar value is that it finishes the surface is a way that brings out all the black, gray and silvery tones that naturally belong to iron, and also prevents it from rusting. This method applies to both wrought iron and sheet iron and is the only thing we know that accomplishes the desired result. The process itself is very simple. After the iron is hammered it should be polished on an emery belt; or if this is not at hand and it is not convenient to borrow the use of one in some thoroughly equipped metal shop, emery

ONE OF THE LITTLE LANTERNS THAT IS FREQUENTLY USED WITH THE SHOWER LIGHTS.

ELECTRIC LANTERN DESIGNED AS A FINIAL TO A NEWEL POST.

cloth—about Number O—may be used in polishing the surface by hand.

Then the iron must be smoked over a forge or in a fireplace, care being taken to avoid heating it to any extent during this process, as the object is merely to smoke it thoroughly. It should then be allowed to cool naturally and the surface rubbed well with a soft cloth dipped in oil. Naturally, the more the iron is polished the brighter it will be, especially in the higher parts of an uneven surface, which take on almost the look of dull silver. After this the piece must be well wiped off so that the oil is thoroughly removed, and the surface lacquered with a special iron lacquer.

To give the copper the deep mellow brownish glow that brings it into such perfect harmony with the fumed oak, the finished piece should be rubbed thoroughly with a soft cloth dipped in powdered pumice stone, and then left to age naturally. If a darker tone is desired, it should be held over a fire or torch and heated until the right color appears. Care should be taken that it is not heated too long, as copper under too great heat is apt to turn black. We use no lacquer on either copper or brass, age and exposure being the only agents required to produce beauty and variety of tone. All our brass work is made of the natural unfinished metal, which has a beautiful greenish tone and a soft dull surface that harmonizes admirably with the natural wood. Like copper, it darkens and mellows with age.

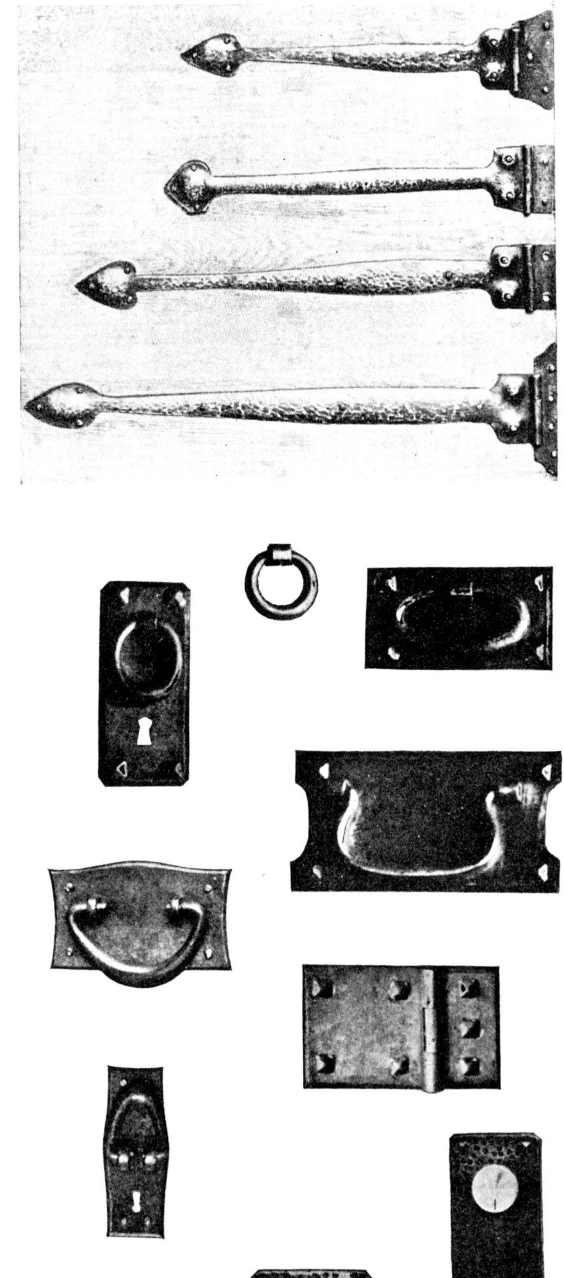

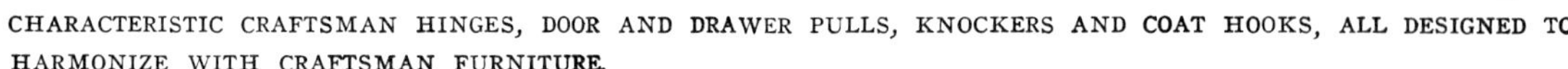

CHARACTERISTIC CRAFTSMAN HINGES, DOOR AND DRAWER PULLS, KNOCKERS AND COAT HOOKS, ALL DESIGNED TO HARMONIZE WITH CRAFTSMAN FURNITURE.

THE KIND OF FABRICS AND NEEDLEWORK THAT HARMONIZE WITH AND COMPLETE THE CRAFTSMAN DECORATIVE SCHEME

WE have traced in this book the development of the Craftsman scheme of building and interior decoration, beginning with the house as a whole and thence working back to an analysis of the different rooms, the wall spaces, struc-

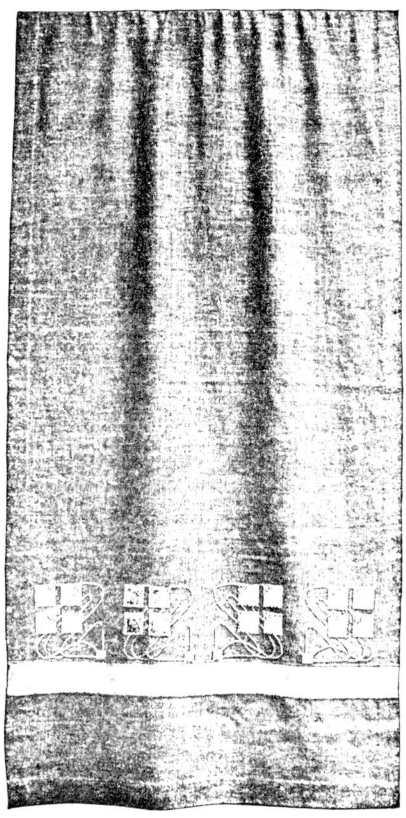

PORTIÈRE OF CRAFTSMAN CANVAS WITH PINE CONE DESIGN IN APPLIQUÉ.

tural features, furnishings and metal work, all of which must be considered separately as essential parts of the complete structure, including the decorative scheme. In doing this we have reversed the process by which we worked out the idea in the first place, for we began ten years ago with the furniture; the metal work followed as a matter of course because it was the next thing needed; then the dressing of leathers to harmonize with the style of the furniture and the wood of which it was made. Then came the finding of suit-

able fabrics and the kind of decoration most in keeping with them, and from all these parts was naturally developed the idea of the Craftsman house as a whole.

At first it was very difficult to find just the right kind of fabric to harmonize with the Craftsman furniture and metal work. It was not so much a question of color, although of course a great deal of the effect depended upon perfect color harmony, as it was a question of the texture and character of the fabric. Silks, plushes and tapestries, in fact delicate and perishable fabrics of all kinds, were utterly out of keeping with Craftsman furniture. What we needed were fabrics that possessed sturdiness and durability; that were made of materials that possessed a certain rugged and straightforward character of fiber, weave and texture,—such a character as

PORTIÈRE OF CRAFTSMAN CANVAS WITH CHECKERBERRY DESIGN IN APPLIQUÉ.

SASH CURTAIN OF TEA-COLORED NET DARNED IN AN OPEN PATTERN WITH SILVER-WHITE FLOSS.

SASH CURTAIN OF CASEMENT LINEN WITH FRETWORK DESIGN IN SOLID DARNED WORK.

would bring them into the same class as the sturdy oak and wrought iron and copper of the other furnishings. Yet they could not be coarse or crude, for that would have taken them as far away from the quality of the furniture on the one side, as plushes and brocades were on the other.

For upholstering the furniture itself we had found leather more satisfying than anything else, especially as by constant experimenting we had succeeded in developing a method of dressing that preserved all the leathery quality in much the same way that we were able to preserve the woody quality of the oak, so that the leather maintained its own sturdy individuality, at the same time possessing a softness and flexibility and a sub-

tlety of coloring that proved wonderfully attractive. This was especially the case with sheepskin, which we finished in all the subtle shades of brown, biscuit, yellow, gray, green, and fawn, but always with the leathery quality predominant under the light surface tone. These leathers accorded so well with the plain oak furniture and metal work that for a time they became almost too popular, for they were used by many people for table covers, portiéres and the like, in rooms where rugged effects were considered desirable. In fact, the fad ran to such lengths that it fortunately wore itself out and leather was allowed to return to its proper uses.

This was made easier by the discovery of certain fabrics that harmonize as completely as leather with the general Craftsman scheme. These are mostly woven of flax left in the natural color or given some one of the nature hues. There are also certain roughly-woven, dull-finished silks that fit into the picture as

SCARF OF HAND-WOVEN LINEN WITH PINE CONE DESIGN IN DARNED WORK.

SASH CURTAIN OF CASEMENT LINEN WITH ANOTHER FORM OF PINE CONE DESIGN DONE IN DARNED WORK.

TABLE SCARF OF UNBLEACHED HAND-WOVEN LINEN WITH DRAGONFLY DESIGN DARNED IN PERSIAN COLORS.

TABLE SCARF WITH GINKGO DESIGN IN APPLIQUÉ OF DEEP LEAF GREEN UPON HOMESPUN LINEN.

well as linen, and for window curtains we use nets and crepes of the same general character. A material that we use more than almost any other for portiéres, pillows, chair cushions,—indeed in all places were stout wearing quality and a certain pleasant unobtrusiveness are required—is a canvas woven of loosely twisted threads of jute and flax and dyed in the piece,—a method which gives an unevenness in color that amounts almost to a two-toned effect because of the way in which the different threads take the dye. This unevenness is increased by the roughness of the texture, which is not unlike that of a firmly woven burlap. The colors of the canvas are delightful. For example, there are three tones of wood brown,—one almost exactly the color of old weather-beaten oak; another that shows a sunny yellowish tone; and a third that comes close to a dark russet. The greens are the foliage hues,—one dark and brownish like rusty pine needles, another a deep leaf-green; the third an intense green like damp grass in the shade; and a fourth a very gray-green with a bluish tinge like the eucalyptus leaf.

Our usual method of decorating this canvas is the application of some bold and simple design in which the solid parts are of linen appliqué in some contrasting shade and the connecting lines are done in heavy outline stitch or couching with linen floss. This sim-

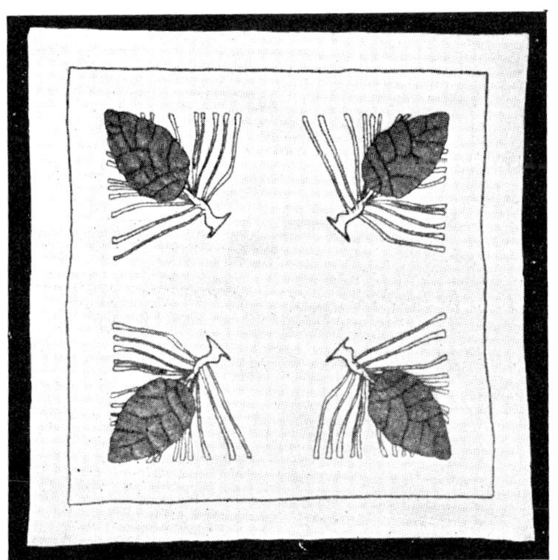

TABLE SCARF OF HOMESPUN LINEN WITH PINE CONE DESIGN IN APPLIQUÉ.

PILLOW COVERED WITH CRAFTSMAN CANVAS AND ORNAMENTED WITH PINE CONE DESIGN IN APPLIQUÉ.

CRAFTSMAN FABRICS AND NEEDLEWORK

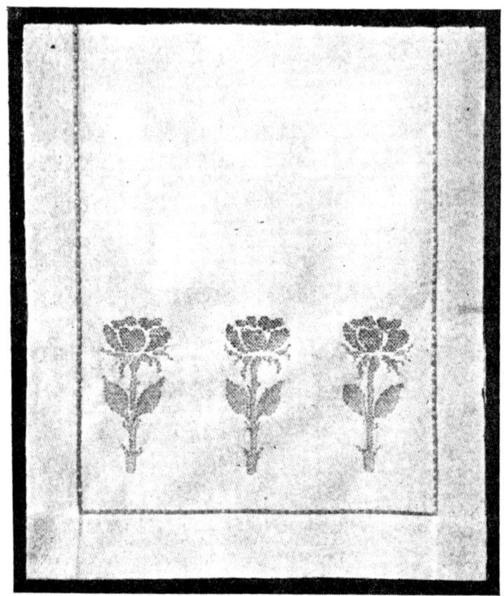

TABLE SCARF FOR A BEDROOM, WITH POPPY DESIGN IN DARNED WORK.

plicity is characteristic of all the Craftsman needlework, which is bold and plain to a degree. We use appliqué in a great many forms, especially for large pieces such as portières, couch covers, pillows and the larger table covers. For scarfs, window curtains and table furnishings of all kinds we are apt to use the simple darning stitch, as this gives a delightful sparkle to any mass of color. For the rest we use the satin stitch very occasionally when a snap of solid color is needed for accenting now and then a bit of plain hem-stitching or drawn work. It is the kind of needlework that any woman can do and, given the power of discrimination and

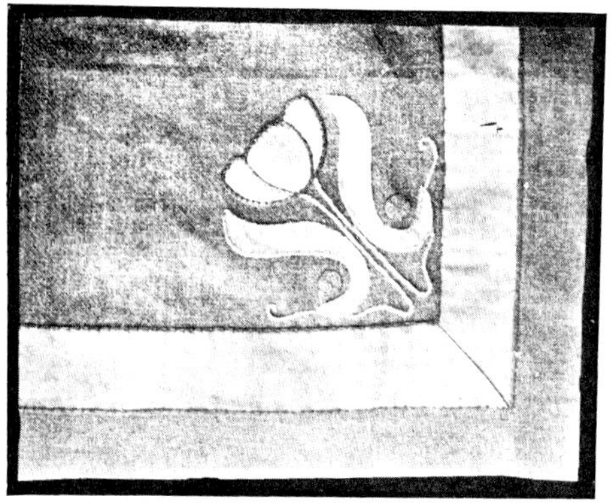

POPPY DESIGN CARRIED OUT IN APPLIQUÉ TO ORNAMENT THE CORNER OF A COUCH COVER.

taste in the selection of materials, designs and color combinations, there is no reason why any woman should not, with comparatively little time and labor, make her home interesting with beautiful and characteristic needlework that is as far removed from the "fancy work" which too often takes the place of it, as any genuine and useful thing is removed from things that are unnecessary.

For scarfs, table squares, luncheon and dinner sets and the like, we find that the most suitable fabrics in connection with the Craftsman furnishings are the linens, mostly in the natural colors and the rougher weaves.

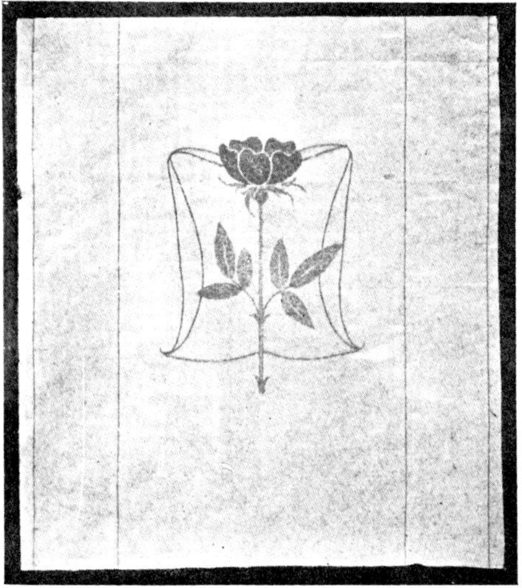

SAME POPPY DESIGN AS APPLIED TO A BEDSPREAD OF HOMESPUN LINEN.

We use hand-woven and homespun linens in many weights and weaves, and a beautiful fabric called Flemish linen, which has a matt finish and is very soft and pliable to the touch. Some of these come in the cream or ivory shades and all of them in the tones of cream gray and warm pale brown natural to the unbleached linen. We find, as a rule, that the finer and more delicate white linens do not belong in a Craftsman room any more than silks, plushes and tapestries in delicate colorings belong with the Craftsman furniture. The whole scheme demands a more robust sort of beauty,—something that primarily exists from use and that fulfils every requirement. The charm that it possesses arises from the completeness with which it answers all these demands and the honesty which allows its natural quality to show.

CABINET WORK FOR HOME WORKERS AND STUDENTS WHO WISH TO LEARN THE FUNDAMENTAL PRINCIPLES OF CONSTRUCTION

IN the brief sketch we have already given of furniture making in this country we made the statement that one of the chief elements of interest in Craftsman furniture is the fact that its construction is so simple and direct and so clearly revealed that any one possessing even a rudimentary knowledge of tools and of drawing and some natural skill of hand could easily make for himself many pieces of furniture in this style. Believing this thoroughly, and also realizing fully the interest that cabinetwork holds for most people and the means it affords of developing the constructive and creative faculties, we have given in THE CRAFTSMAN a number of designs solely for the benefit of home workers. For a year or two we published, in connection with these designs, full working drawings and also mill bills for the necessary lumber; but we were forced to abandon that on account of lack of space and to give only the drawings showing the finished pieces, for which the working drawings and mill bills were easily obtainable upon application.

We illustrate here a number of these designs, most of which are for pieces that are fairly easy to make and that have a definite use as household furnishings. While the designs of course show the exact models of the pieces they represent, we intend them to have also a suggestive value and to stimulate thought and experiment along the lines of designing and making plain substantial furniture. It has been proven beyond question that the most powerful stimulus to well-defined constructive thought is found in the direction of the mind to some form of creative work. Therefore if a man or a boy has any aptitude along these lines, it is a foregone conclusion that he will not have

FIGURE ONE.—SQUARE TABOURET.

made many pieces after given models before he begins to think for himself and to make or modify designs to meet his own demands and to afford an opportunity for working out his own problems. Furthermore, as his experience grows, he will naturally discover new ways of doing things that may be better for him to follow than any of the stereotyped rules. We approve thoroughly of the freedom of spirit that leads to such experimenting, for, although we originated the Craftsman furniture, it is just such interest and work on the part of other people that will ultimately develop it into a national style. One warning, however, we would like to give to all amateur workers: that is, that one's own whims must no more be followed than the whims of other people. We will find plenty of interest and occupation in making things that are actually needed and plenty of exercise for all our creative power in designing them to fulfil as adequately as possible the purpose for which they are intended. So long as this is done there is no danger of the work degenerating into a fad; instead, it is likely not only to give much pleasure and profit to individuals, but to grow until the whole nation once more reaps the benefit that comes from the intelligent exercise of the creative powers in some interesting form of handicraft.

Every one knows the relief to brain workers and to professional men that is found in this kind of work. It not only affords a wholesome change of occupation but brings into play a different set of faculties and so proves both restful and stimulating. A professional or business man who can find relief from his regular work in some such pursuit,

FIGURE TWO.—A ROUND TABOURET.

CABINET WORK FOR HOME WORKERS

FIGURE THREE.—HALL BENCH WITH CHEST.

peals to him instead of being taught sound principles of design and construction and so guided by a competent worker that all his own work is based upon these principles and is thoroughly done. If the work is m e r e l y regarded as play, the theoretical attitude toward the expression of individuality is all right; but if it is regarded as a preparation for the serious business of later life, the result shows that it unfits the student for real work in just such measure as he shows an aptitude for play work.

which he takes up as a recreation, does better work in his own vocation because he is a healthier and better balanced man and his interest in his home grows more vivid and personal with every article of furniture that he makes with his own hands and according to his own ideas.

As for the means of education afforded by this kind of work, we have no better proof than is shown by the widespread belief in the efficacy of manual training in our public schools, although to a practical craftsman there would seem to be plenty of room for improvement, both as to methods of teaching and the quality of workmanship that is required from the students. W h e r e manual training is taken up purely on account of the mental development it affords, there is a tendency to make it entirely academic. The teachers for the most part rely almost wholly upon theory and have very little practical knowledge of the t h i n g they teach. The result is that a boy is encouraged to "express his own individuality" in designing and making the thing that ap-

The introduction of the Craftsman style has practically revolutionized manual training in our public schools, because it has placed at the disposal of the teachers designs of such simplicity and clearness of construction that the work of teaching has been made much easier and the field of manual training has been greatly broadened. Before the introduction of Craftsman furniture, manual training in the schools rested chiefly upon sloyd, which was confined to the making of small articles entirely for the sake of the mental development afforded by the intelligent use of the hands. Now, however, the students of manual training are learning to

FIGURE FOUR.—CHILD'S OPEN BOOKCASE.

FIGURE FIVE.—CHESS OR CHECKER TABLE.

make furniture after such models as we show here and the very necessary element of usefulness is added to the things they make. The only difficulty is that the craft itself is not well enough understood by the teachers to be imparted to the students in such a way that they derive any permanent benefit from it. The teaching is, as we have said, largely theoretical and the object of the whole training is mental development along general lines rather than the moral development that comes from learning to do useful work thoroughly and well. As cabinetwork is handled in the manual training departments of the schools, it is distinctly a side issue, and exhibitions of the work to which public attention is frequently invited show ambitious pieces of furniture that are wrongly proportioned, badly put together and finished in a slovenly way, thus producing exactly the opposite effect upon the pupil from what is intended. If the State or municipal authorities would see to it that manual training in the form of wood-working of all kinds, and especially the making of furniture, were placed under the charge of thoroughly skilled craftsmen who understood and were able to teach all the principles of construction, the moral and educational effect of such work would be almost incalculable.

In order to make the training of any real value, it is absolutely necessary that the student begin simultaneously with mechanical drawing and the application of its principles

FIGURE SEVEN.—PORTABLE CABINET FOR WRITING TABLE.

to his work as he goes along. If he began with simple models to which could be applied the elementary lessons in mechanical drawing, the laying out of plans, the reading of detail drawings and the like, and would also afford a chance to demonstrate lessons in the use of the square, the level, the saw and the plane;—a good foundation would be laid not only for the understanding of right principles of construction but for the accurate use of tools. A boy trained in this way would be able in future years to put his knowledge to almost any use that was needed. Instead of this the students endeavor to make something that is interesting and that shows well at home or in an exhibition. In fact, the situation now is very much as it would be if a student of music were to take two or three lessons in the rudiments and then endeavor to play a more or less elaborate composition. There is no question as to the benefit that boys, and girls too, derive from being taught to work with their hands; but it is better not to teach them at all than to give them the wrong teaching. No one expects a schoolboy or an amateur worker of any age to make elaborate furniture that would equal similar pieces made by a trained cabinetmaker. But if the student be taught to make small and simple things and to make each one so that it would pass muster anywhere, he learns from the start the fundamental principles of design and proportion and so comes naturally to understand what is meant by thorough workmanship.

There is no objection to any worker, however inexperienced, attempting to express his own

FIGURE SIX.—PIANO BENCH, STRONGLY MADE WITH SOLID ENDS.

FIGURE EIGHT.—CHILD'S WRITING DESK.

and construction as carefully as he would be grounded in mathematics or classical literature, he might safely be trusted to produce something that would express his own individuality, for then, if ever, he would have developed an individuality that was worth while. And this principle applies as well to amateur workers of all kinds as it does to the students in the public schools, for it is the basis of all work that is worthy to endure.

One great advantage of taking up cabinetmaking at home as well as in the schools, is that it could be made not only a means of amusement or mental development to the individual, but could be expanded into a home or neighborhood handicraft that might be carried on in connection with small farming, upon a basis that would insure a reasonable financial success. Handicrafts, as practiced by individual arts and crafts workers in the studio, do not afford a sufficient living to craft workers as a class, but that is largely because these very principles of sound construction and thorough workmanship are not always observed or even comprehended, so that it is difficult for the individual worker to produce anything that has a definite and permanent commercial value. This kind of furniture, on the contrary, has a very well defined and thoroughly established commercial value, as our own experience has proven; and yet it is so simple in design and construction that it can be made at home or on the farm during the idle months of winter or by a group of

individuality, but the natural thing would be for him to express it in more or less primitive forms of construction that are, so far as they go, correct, instead of attempting something that, when it is finished, is all wrong because the student has not understood what he was about. Unquestionably there are certain principles and rules as to design, proportion and form that are as fundamental in their nature as are the tables of addition, subtraction, division and multiplication, with relation to mathematics, or as the alphabet is as a basis to literature, but they are not yet formulated for general use. The trained worker learns these things by experience and comes to have a sort of sixth sense with regard to their application, but this takes strong direct thinking, keen observation and the power of initiative that is possessed only by the very exceptional and highly skilled workman.

Nevertheless it surely is as easy to begin work in the right way as in the wrong way. It would be better if all our teaching of manual training were based upon some text book carefully compiled by a master workman and kept within certain well defined limits. After the student had thoroughly learned all that lay within these limits and was grounded in the principles of design

FIGURE NINE.—BRIDE'S CHEST.

FIGURE TEN.—BOOK CABINET.

have much more individual interest because of this very feature, for then it would be possible to select certain pieces of wood for special uses and to develop to the utmost all the natural qualities of color and grain that might prove interesting when rightly used and in the right place. It is by these very methods and under similar conditions that the Japanese have gained such world-wide fame as discriminating users of very simple and inexpensive woods. A Japanese regards a piece of wood as he might a picture and his one idea is to do something with it that will show it to the very best advantage, as well as gain from it the utmost measure of usefulness.

Among the cabinet woods native to this country and easily obtained are white oak, brown ash, rock elm, birch, beech and maple. Chestnut, cypress, pine, redwood and gumwood, while all excellent for interior trim,

workers in a village,—in fact under almost any conditions where it would seem advantageous to do such work, especially under the guidance of a competent cabinetmaker.

Whether regarded as one of the forms of a profitable handicraft that might be depended upon as a means of support,—or at least of adding to the income obtained from a small farm,—or whether regarded merely as a means of recreation for a busy man during his leisure hours at home, cabinetmaking is likely to prove a most interesting pursuit. One distinct advantage is that furniture made in this way, if well done, would be better than any that could possibly be made in a factory, because the work would naturally be more carefully done. Also the interest that attaches to the right use of wood could be developed to a much greater degree than is possible where the work is done on a large scale, because judgment and discrimination could be applied to the selection of lumber that is without any special market value according to commercial standards, but that has in it certain flaws and irregularities that make it far more interesting than the costlier lumber necessary for purely commercial work. This one item would be a great advantage as lumber grows scarcer and harder to obtain. Also, the furniture itself would

FIGURE ELEVEN.—BOOKCASE WITH ADJUSTABLE SHELVES.

FIGURE TWELVE.—SMALL STAND FOR USE IN A BEDROOM.

are not hard enough to give satisfactory results when used for the making of furniture. Of those first mentioned, white oak is unquestionably the best for cabinetmaking and, indeed, it is a wood as well suited to the Craftsman style of furniture as the Spanish mahogany was to the French, English and Colonial furniture of the eighteenth century. Spanish mahogany is very rare now and the modern mahogany, or baywood, is very little harder than whitewood and so cannot be considered particularly desirable as a cabinet wood. The old mahogany was a hard, close-grained, fine-textured wood that lent itself naturally to the slender lines, graceful curves and delicate modeling of the eighteenth century styles. In addition to this the wood itself was so treated as to ripen to the utmost the quality of rich and mellow coloring, which was one of its distinctive characteristics. The boards were kept for months, and some of them for years, in the courtyards of the cabinet shops, where sun and rain could give them the mellowness of age. Then the finished pieces were treated with linseed oil and again put out into the sunshine to oxidize, this process being repeated until the wood gained just the required depth of color and perfection of finish. The slowness of this process and the care and skill required to produce the results that were aimed at makes fine mahogany furniture almost an impossibility today, except to the craftsman who may be able to afford selected pieces of this rare and almost extinct wood, and who has sufficient leisure and love of the work to treat it according to the methods of the old cabinetmakers. Even then it is not suitable for the plain massive furniture that

we show here as models for home workers. The severely plain structural forms that we are considering now demand a wood of strong fiber and markings, rich in color, and possessing a sturdy friendly quality that seems to invite use and wear. The strong straight lines and plain surfaces of the furniture follow and emphasize the grain and growth of the wood, drawing attention to instead of destroying the natural character that belonged to the growing tree. As the use of oak would naturally demand a form that is strong and primitive, the harmony that exists between the form and construction of the furniture and the wood of which it is made is complete and satisfying.

We will then assume that oak is the wood that would naturally be selected by the home cabinetmaker and for large surfaces such as table-tops and large panels, quarter-sawn oak is deemed preferable to plain-sawn, as the first method, which makes the cut parallel with the medullary rays that form the peculiar wavy lines seen in quarter-sawn oak, not only brings out all the natural beauty of the markings, but makes the wood structurally stronger, finer in grain and less liable to check and warp than when it is straight-sawn. Care should then be taken to see that the wood is thoroughly dried, otherwise the best work might easily be ruined by the checking, warping, or splitting of the lumber. Quarter-sawn oak is the hardest of all woods to dry and requires the longest time, so that it would hardly be advisable for the amateur cabinetmaker to attempt to use other than selected kiln-dried wood that is ready for the saw and plane.

FIGURE THIRTEEN.—ROUND TABLE.

FIGURE FOURTEEN.—WRITING DESK WITH WILLOW
WASTE BASKET.

The work of construction must all be done before the wood is given its final finish; but in this connection we will outline briefly the best method of finishing oak, as the sturdy wooden quality of the furniture depends entirely upon the ability of the worker to treat the wood so that there is little evidence of an applied finish. Oak should be ripened as the old mahogany was ripened by oil and sunshine, and this can be done only by a process that, without altering or disguising the nature of the wood, gives it the appearance of having been mellowed by age and use. This process is merely fuming with ammonia, which has a certain affinity with the tannic acid that exists in the wood, and it is the only one known to us that acts upon the glossy hard rays as well as the softer parts of the wood, coloring all together in an even tone so that the

figure is marked only by its difference in texture. This result is not so good when stains are used instead of fuming, as staining leaves the soft part of the wood dark and the markings light and prominent.

The fuming is not an especially difficult process, but it requires a good deal of care, for the piece must be put into an air-tight box or closet, on the floor of which has been placed shallow dishes containing aqua ammonia (26 per cent). The length of time required to fume oak to a good color depends largely upon the tightness of the compartment, but as a rule forty-eight hours is enough. When fuming is not practicable, as in the case of a piece too large for any available compartment or one that is built into the room, a fairly good result may be obtained by applying the strong ammonia directly to the wood with a sponge or brush. In either case the wood must be in its natural condition when treated, as any previous application of oil or stain would keep the ammonia from taking effect. After the wood so treated is thoroughly dry from the first application it should be sandpapered carefully with fine sandpaper, then a second coat of ammonia applied, followed by a second careful sandpapering.

Some pieces fume much darker than others, according to the amount of tannin left free to attract the ammonia after the wood has

FIGURE FIFTEEN.—TABLE DESK.

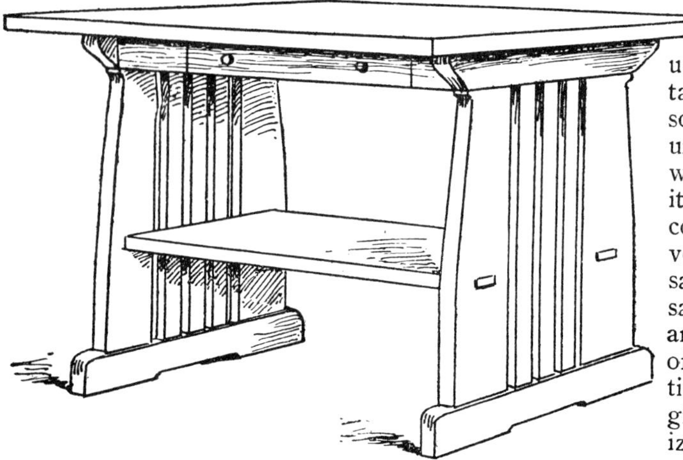

FIGURE SIXTEEN.—LIBRARY TABLE.

ened by the addition of a small quantity of the stain used in touching up. Care must be taken, however, to carry on the color so lightly that it will not grow muddy under the brush of an inexperienced worker. The danger of this makes it often more advisable to apply two coats of lacquer, each containing a very little color. If this is done, sandpaper each coat with very fine sandpaper after it is thoroughly dried and then apply one or more coats of prepared floor wax. These directions, if carefully followed, should give the same effects that characterize the Craftsman furniture.

Sometimes a home cabinetworker does not find it practicable or desirable to fume the oak. In such a case there are a number of good stains on the market that could be used on oak as well as on other woods.

Oak and chestnut alone are susceptible to the action of ammonia fumes, but in other ways the oak, chestnut, ash and elm come into one class as regards treatment, for the reason that they all have a strong, well-defined grain and are so alike in nature that they are affected in much the same way by the same method of finishing. For any one of these woods a water stain should never be used, as it raises the grain to such an extent that in sandpapering to make it smooth again, the color is sanded off with the grain, leaving an unevenly stained and very unpleasant surface. The most satisfactory method we know, especially for workers who have had but little experience, is to use a small amount of color carried on in very thin

been kiln-dried. Where any sap wood has been left on, that part will be found unaffected by the fumes. There is apt also to be a slight difference in tone when the piece is not all made from the same log, because some trees contain more tannic acid than others. To meet these conditions it is necessary to make a "touch-up" to even the color. This is done by mixing a brown aniline dye (that will dissolve in alcohol) with German lacquer, commonly known as "banana liquid." The mixture may be thinned with wood alcohol to the right consistency before using. In touching up the lighter portions of the wood the stain may be smoothly blended with the darker tint of the perfectly fumed parts, by rubbing along the line where they join with a piece of soft dry cheese-cloth, closely following the brush. If the stain should dry too fast and the color is left uneven, dampen the cloth very slightly with alcohol. After fuming, sandpapering and touching up a piece of furniture, apply a coat of lacquer, made of one-third white shellac and two-thirds German lacquer. If the fuming process has resulted in a shade dark enough to be satisfactory, this lacquer may be applied clear; if not, it may be dark-

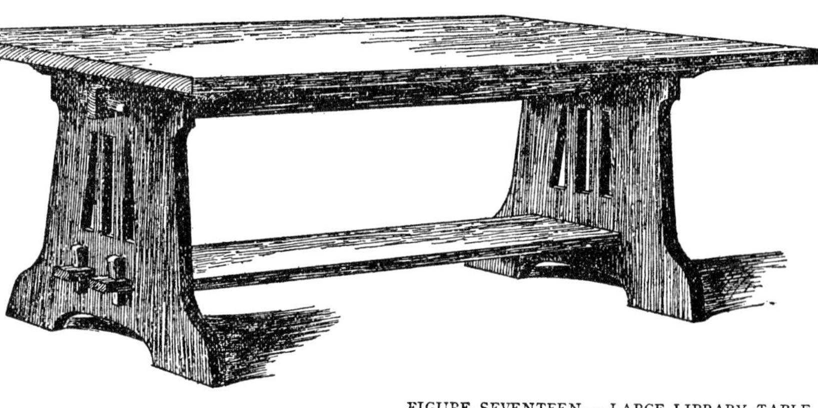

FIGURE SEVENTEEN.—LARGE LIBRARY TABLE.

shellac. If the commercial cut shellac is used it should be reduced with alcohol in the proportion of one part of shellac to three of alcohol. This is because shellac, as it is ordinarily cut for commercial purposes, is mixed in the proportion of four pounds to a gallon of alcohol, so that in order to make it thin enough it is necessary to add sufficient alcohol to obtain a mixture of one pound of shellac to a gallon of alcohol. If the worker does his own cutting he will naturally use the proportion last mentioned,—

FIGURE NINETEEN.—PLATE RACK TO BE PLACED OVER A SIDEBOARD.

one pound of shellac to a gallon of alcohol. When the piece is ready for the final finish, apply a coat of thin shellac, adding a little color if necessary; sandpaper carefully and then apply one or more coats of liquid wax. These directions are entirely for the use of home workers. The method we use in the Craftsman Workshops differs in many ways, for we naturally have much greater facilities for obtaining any desired effect than would be possible with the equipment of a home worker.

For lighter pieces of furniture suitable for a bedroom or a woman's sitting room, where dainty effects are desirable, we find maple the most satisfactory, in both color and texture, of our native woods, for the reason that it is hard enough to be used for all kinds of furniture. Gumwood is equally beautiful, but is not hard enough for chairs. For built-in furniture, however, and for tables, dressers and the like, gumwood is one of the most beautiful woods we have, as it takes on a soft, satin-like texture with variable color effects not unlike those seen in the finest Circassian walnut. We find that the best effect in both maple and gumwood is obtained by treating the wood with a solution of iron-rust made by throwing iron filings or any small pieces of iron into acid vinegar or a weak solution of acetic acid. After forty-eight hours the solution is drained off and diluted with water until the desired color is obtained. The wood is merely brushed over with this solution,—wetting it thoroughly,—and left to dry. This is a process that requires much experimenting with small pieces of wood before attempting to treat the furniture, as the color does not show until the application is completely dry. By this treatment maple is given a beautiful tone of pale silvery gray and the gumwood takes on a soft pale grayish brown, both of which colors harmonize admirably with dull blue, old rose, straw color, or any of the more delicate shades so often used in furnishing a bedroom or a woman's sitting room.

As to the actual construction of the pieces shown here, it is in most cases very simple. By a careful study of the different models it will be noted that the only attempt at decoration lies in the emphasizing of the actual structural features, such as posts, panels, tenons with or without the key, the dovetail joint and the key as

FIGURE EIGHTEEN.—SMALL SIDEBOARD.

213

it is used to strengthen and emphasize the joining of two boards. For the rest, the beauty of each piece depends wholly upon the care with which the wood is selected, the proportions and workmanship of the piece, and the attention that is given to the delicate details of construction and to the finish of the wood.

In Figures 1 and 2 we illustrate two of the simplest models we have ever offered for the use of home cabinetworkers. They are two designs for small tabourets and were selected to illustrate the first article on home training in cabinetwork, published in THE CRAFTSMAN in April, 1905. Therefore from the point of view of their precedence in the series, no less than their fitness as models for the beginner, they have been chosen to head the illustrations for this article. In the case of both of them the construction shows for itself. The tenons of the legs are visible through the top of the table, where they are firmly wedged and then planed flush with the top. This not only strengthens the table very considerably, but the difference in the grain of the wood gives the effect of four small square inlays in each table top. Also it is well to note that, in cutting the mortises for the stretchers of the square tabouret, there is half an inch difference in the heights of the two stretchers. A dowel pin three-eighths of an inch in diameter

FIGURE TWENTY-ONE.—SMALL LETTER AND PAPER FILE.

runs all the way through the legs and holds firm the tenons of the stretchers, making it practically impossible for the table to rack apart. These pins are planed off flush with the sides of the legs.

Figures 3 and 6 illustrate companion pieces, the first being a hall bench and the second a piano bench. Both are simple to a degree, yet the proportions are so contrived that the effect of each is individual and decorative. The outward slope of the solid end pieces gives an appearance of great strength that does full justice to the real strength of both benches. The severity of these end pieces is rather lightened by the curved opening at the bottom and by the openings at the top, meant in each case for convenience in moving the bench. These openings, with the slight projections of the tenons at the ends, form the only decoration. In the case of the hall bench, a shallow box takes the place of the curved brace that appears under the seat in the piano bench. This box can be used to hold all sorts of things that ordinarily accumulate in the hall and the hinged seat lifts like a lid over it. The bench can be made in any desired length to fit any wall space without interfering with its construction or proportions.

Figure 4 shows a small open bookcase that is intended for the use of children. All housewives know that one of the greatest difficulties in keeping a tidy nursery often arises because there is no place where children can easily put things away themselves. Closet doors are hard to open and the shelves too high to be of use, while shelves and brackets are usually purposely out of reach and the nursery table is apt to be full. This little bookcase is planned especially to meet just such a nursery problem. There are no doors and the shelves are broad and

FIGURE TWENTY.—COMBINATION TABLE AND ENCYCLOPEDIA BOOKCASE.

low enough to be within the reach of very little children. The shelves are not adjustable but are put in stoutly with tenon and key so that they are never out of place and never need attention.

Figure 5 shows a small table that would prove a convenient piece of furniture in a household where either chess or checkers happens to be a favorite game. The legs are slightly tapering, sloped outward and are made firm with bracket supports so that the usual cross supports below, which would interfere with the comfort of the players sitting at the table, are not needed. The rails under the top are tenoned to the legs. In a case like this, where two or more rails meet with the ends opposite each other, short tenons must be used with two dowel pins in each one to hold it in place. These pins are placed near the edge of the table legs so that they may not interfere with the tenoning of the side rails. It is a good plan to dowel the bracket supports first to the legs and then to the top of the table, in addition to the glue which holds them in place. The small drawer is made in the regular way, being hung from the top instead of running on a center guide as do most of the wider drawers in Craftsman furniture. The checks on the table top may be burned into the wood or a dye or stain may be used for the dark checks.

Figure 7 shows a small portable cabinet that may be placed on the top of any writing table. It is provided with little compartments which are protected by doors with flat key locks and with a shelf and pigeon-holes for papers and books. The piece is perfectly plain except for the slight decorative touch given by the dove-tailing at the end, but if the wood is well chosen and the cabinet carefully finished, it will be found an attractive as well as a convenient bit of furniture.

Figure 8 illustrates a child's desk, the making

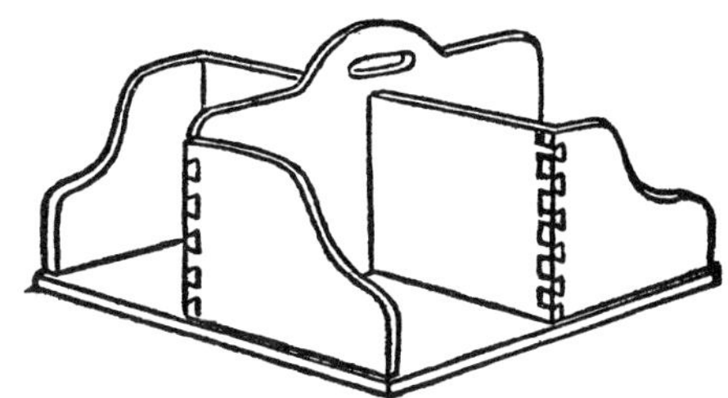

FIGURE TWENTY-TWO.—SMALL REVOLVING BOOK-RACK.

of which would be an especially pleasant piece of work for the home craftsman, because there is no article of miniature furniture which affords the children so much delight as a desk where they can work like grown-up folks and have pads and pencils never to be loaned or lost. This little desk is so simple that the small members of the family might even help to make it and so gain some understanding of the pleasure of making their own belongings. The construction has the same general features that have already been described and the only touch of decoration is the projection of the two back posts above the small upper shelf.

Figure 9 suggests a useful and desirable present for a bride, for it is a cedar-lined chest

FIGURE TWENTY-THREE.—COMBINATION BOOKCASE AND CUPBOARD.

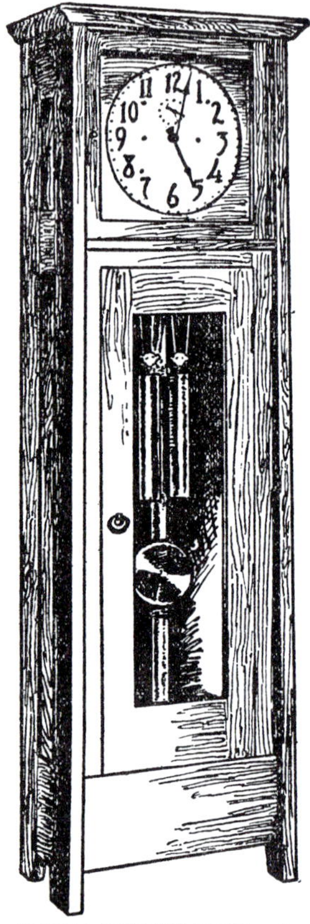

FIGURE TWENTY-FOUR.—A HALL CLOCK.

intended for the storing of linen and clothing,—just the same sort of chest as the German maidens use for storing away the linen they weave during their girlhood. In making the chest the legs are first built up, then the front and back fastened in; the ends and bottom are put in at the same time, fitting in grooves. The top is simply made, with two panels divided by a broad stile which affords support for the iron strap-hinge that extends down the side to be fastened with hasp and padlock. The inside of the chest is lined with cedar boards, so desirable for their pleasant aromatic odor and for their moth - preventing properties. This lining should be put in after the chest is made. The iron work can be made by any blacksmith from the drawing, or even made at home if the amateur cabinetworker also possesses a forge.

Figure 10 shows a book cabinet which would be convenient in a workroom, where it might stand near the desk or table of the worker and provide a place for the few books of reference that are in constant use, as well as for papers, drawings and so forth, that might otherwise be mislaid or scattered in confusion about the room. The cabinet is easy to make and is very satisfying in line and proportion. The shelf that covers half the top offers room for a small paper rack or any of the many things that have to be within reach and yet not in the way.

Figure 11 gives a model for a bookcase having two drawers below for papers or magazines and three adjustable shelves that can be moved to any height simply by changing the position of the pegs that support the shelves. If the books are small, an additional shelf might be put in if required. The frame of the bookcase is left plain, the smooth surface of the sides being broken only by the slightly projecting tenons at the top and bottom. The edges of these tenons are chamfered off and carefully sandpapered so that they have a smooth rounded look. Inside the ends of the bookcase holes about half an inch in diameter are bored about halfway through the thickness of the plank, affording places for the pegs that hold the adjustable shelves.

Figure 12 shows a small table primarily chosen for use in a bedroom, to stand near the bed and hold a lamp or candle and one or two books, but it is convenient in any place where a small stand is needed. The top of the back is to be doweled in place with three half-inch dowel pins and the top itself is fastened to the sides by table fasteners placed under a wide overhang. The drawers should be dovetailed together at the corners and all edges slightly softened by careful sandpapering.

Figure 13. The round table shown here embodies in its construction the same general features as the large square library table shown in Figure 17, only modified to such a degree that the effect is light rather than massive. The braces, top and bottom, are crossed and the four legs are wide and flat, with openings following the lines of the outside. The tenons, which have a bold projection and are fastened with wooden keys, are used as a distinctively decorative feature.

Figure 14 gives a very good idea of a desk which looks hard to make but is not so difficult as might appear at the first glance. The lid can be made first, then the sides and shelves carefully fitted and a quarter-inch iron pin inserted between the sides and the lid so that all

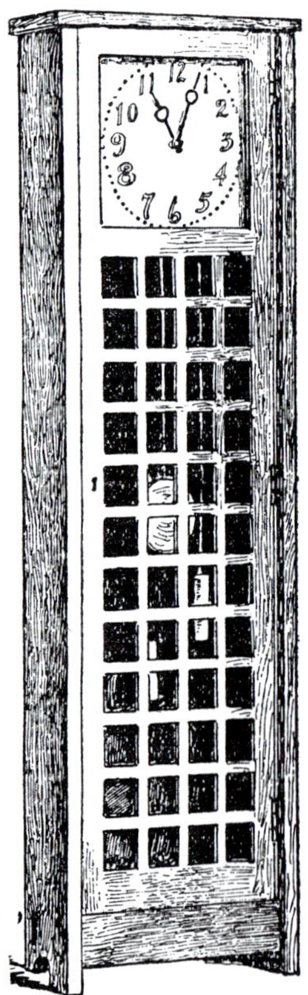

FIGURE TWENTY-FIVE.— A HALL CLOCK.

FIGURE TWENTY-SIX.—CHILD'S HIGH CHAIR.

are fastened together at once. Then the back is put in and is held in place by small blued oval - headed screws. After this the letter a n d blotter rack ˙may be sprung into place and, with a little button at the top under which is a leather washer, the desk is complete except for the basket, which should be woven of willow withes to fit the shelf.

Figure 15 shows a simple desk or writing table for the library or living room. The two little upper drawers with the letter file between give a very convenient arrangement for stowing away letters, writing paper, etc. This is a piece which might easily be made crude and heavy, by just a little awkwardness in getting the right proportions and lack of skill in the use of tools; but if carefully made and well finished it possesses a sturdy attractiveness that is very interesting.

Figure 16. This design for a library table should not be attempted until experience in woodwork has taught the worker how to use his tools and materials well. Everything depends upon care both in construction and finishing and especial attention should be given toward maintaining in their integrity all the lines and proportions, as these details have everything to do with making or marring the design. The end pieces, while massive in effect, are relieved from over-heaviness by the use of slats and the shaping of the broad strips on the outside. The top of the table is fastened firmly with table irons so that it is quite solid. Where the shelf tenons come through the end pieces there is a

projection of three-sixteenths of an inch and the edges are chamfered off to give a smooth rounded effect. The tenon itself should be wedged and glued so that it cannot be pulled out. The dovetailing on the drawer may need a little practice before it is successfully executed, but if it is well done it will be a satisfactory evidence of the cleverness of the worker.

Figure 17 shows a large library table that is practically a companion piece to the round table illustrated in Figure 13. In this case, however, the natural massiveness of the construction is emphasized rather than modified, although the severity of the solid ends is softened by the curved lines and open spaces which serve to take away all appearance of clumsiness. The projecting tenons and keys form a suitable structural decoration and add to the strength of the piece. A strong brace just beneath the top keeps the ends firm while the lower shelf acts as another brace.

Figure 18. The lines and proportions of this small sideboard make it an unusually satisfying piece for the home worker to try his skill on because, if it is well made, it is a piece

FIGURE TWENTY-SEVEN.—PORTABLE SCREEN.

FIGURE TWENTY-EIGHT.—RUSTIC BENCH WITH SLAB TOP.

of furniture that would add much to the beauty of a dining room. The construction, though on a larger scale and in some ways more complicated than in any of the preceding pieces, is no more difficult and no trouble will be found in putting it together. The back is to be screwed into place and is put on last. The top can be doweled on or fastened with table irons. The latter will be safer if there is any doubt as to the thorough seasoning of the wood, as the irons will admit of a slight shrinkage or swelling without cracking the wood. All the edges should be slightly softened with sandpaper just before the finish is applied.

Figure 19. This plate rack is meant to be hung by chains, cords, or heavy picture wire just above the sideboard, although it also serves as a stein rack for a den. The construction speaks for itself and is so simple that nothing need be said about it except that the brackets are fastened with screws from the back. If chains are used to hang it from the rail above, it would be better to have them fairly heavy. Plain round link chains can be bought ready made, together with the hooks, or they can be made to order by any blacksmith.

Figure 20 shows a combination table and encyclopedia bookcase designed especially for the student who wishes to have his reference books near at hand. It is meant to hold a complete set of books, with additional space for a dictionary. The plans are so simple that they can be understood and applied by a beginner in cabinetwork and the usefulness of the piece is such as to make it one of the most interesting models we have ever designed for the use of home workers.

Figures 21 and 22 show two most convenient little pieces for a library table. The first is a small letter file with four compartments for note paper, envelopes and letters, making it very useful for the home bookkeeper. The second is a small revolving book rack, made in the form of a swastika, which revolves upon a flat round stand that raises it about an inch from the table. It is meant to hold small books that are needed for constant reference. Both these pieces show to the best advantage the decorative use of the dovetail as a joint. This bit of structural decoration is a favorite with us because we consider the hand-made dovetail to be one of the most interesting structural features used in joinery, as well as the strongest joint. This, of course, applies only to pieces where the strength of the structure depends upon the strength of the corner, for it is purely a corner joint. For example, in the case of this little book-rack the use of the dovetail is almost inevitable, for without it the corners would not only be less perfectly

FIGURE TWENTY-NINE.—RUSTIC TABLE THAT CAN BE TAKEN APART AT WILL.

218

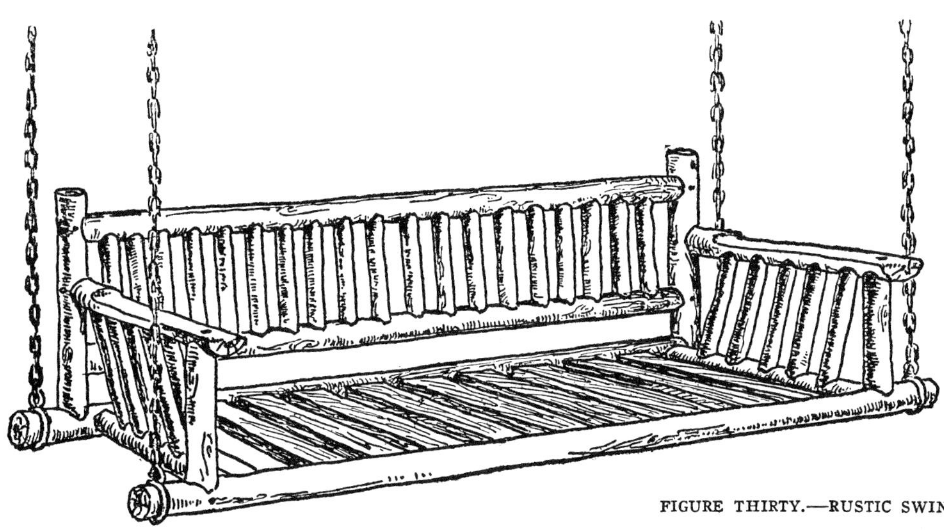

FIGURE THIRTY.—RUSTIC SWING SEAT.

joined as regards strength, but the piece would lose its greatest claim to structural interest.

Figure 23 shows a combination bookcase and cupboard with an open shelf in the middle for such books as are most used. The sides have small-paned glass doors and are shelved for books; the central cupboard with the wooden doors is meant to hold papers, magazines and the like.

Figures 24 and 25 show two hall clocks, of the type usually known as the "grandfather's clock." Given a moderate skill in the handling of tools, the home worker can easily make a clock that will prove a quaint and satisfactory bit of furnishing and will have all the charm of an individual piece of handicraft made for the place it is to fill. Oak is the most appropriate wood for the cases of both these clocks, and the construction is very simple. The face may be made of wood with the figures burned in, or of a twelve-inch plate of brass with figures of copper. If the latter is used, holes should be drilled in the plate to receive the pins which rivet on the figures. These pins are simply bent over after the figures are in place. In both cases the door at the back should have a silk panel in it so that the sound may easily pass through.

Figure 26 shows a child's high-chair designed in the typical Craftsman style. In building this chair put everything together except the arms and when the glue is dry the arm dowels are fitted and the back ones shoved into place. Then by pressure the front will spring into its proper position. All the dowels should be well glued. Care should be used in the joining of the seat rails and it should also be noted that three-eighths of an inch is cut from the bottom of the back post after the chair is put to-

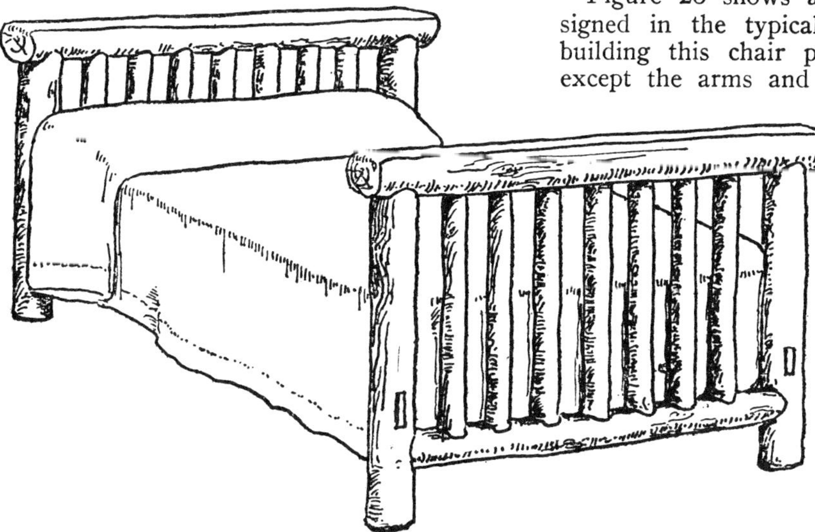

FIGURE THIRTY-ONE.—RUSTIC BED FOR LOG CABIN OR MOUNTAIN CAMP.

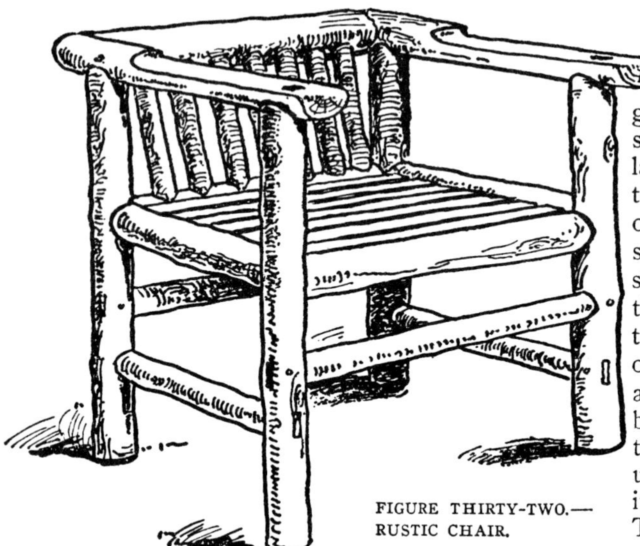

FIGURE THIRTY-TWO.—
RUSTIC CHAIR.

gether. This makes a little slant back to the seat and gives a comfortable position to the sitter. The back slats of the chair are slightly curved—a thing that can be done by thoroughly wetting or steaming the wood and pressing it into shape and then allowing it to dry. The arms of the adjustable tray are cut from a single piece of wood and the back ends are splined by sawing straight in to a point beyond the curve and inserting in the opening made by the saw a piece of wood cut with the grain and well glued. This device gives strength to a point that otherwise would be very weak.

Figure 27 shows a screen which is very easy to make, yet most decorative, owing to the proportion of the leaves, the curving of the top and the use of keys to hold together the broad V-jointed boards of the lower part. The upper part may be of silk, leather, or any material that is preferred.

Figures 28 and 29 show a rustic bench and table

meant for a log cabin or mountain camp. The legs of the bench are made of small logs which are hewn or planed at four angles, leaving the round surface and the wane, so that the piece has in it some of the irregularity of the trunk of the growing tree. The top of the bench is made of a split log planed only at the upper side, the under side being stripped of its bark and left in the natural shape. The horses for the table are made in the same way as the legs of the bench. The table top is in two pieces, the wide thick planks of which it is made being finished as carefully as for any well-made table. These table boards are locked together underneath so that there is no danger of their parting when in use and they can easily be taken apart when it is necessary to move or set aside the table. The great convenience of this table is that it can be taken to pieces and used anywhere, indoors or out.

Figures 30, 31, 32 and 33 show some substantial pieces of rustic furniture designed for country or camp life or for outdoor use. The first is a swinging seat for the veranda or lawn; the second, a bedstead for use in a log cabin or camp; the third is a rustic chair and the fourth a rustic couch for outdoor use. The value of this rustic furniture is not wholly that it is durable and capable of weathering sun and rain alike, but that it makes a special appeal to the amateur carpenter, as its rough exterior hides defects in joining and there is not the special need of well seasoned and carefully prepared lumber that is so essential to the success of the finer pieces.

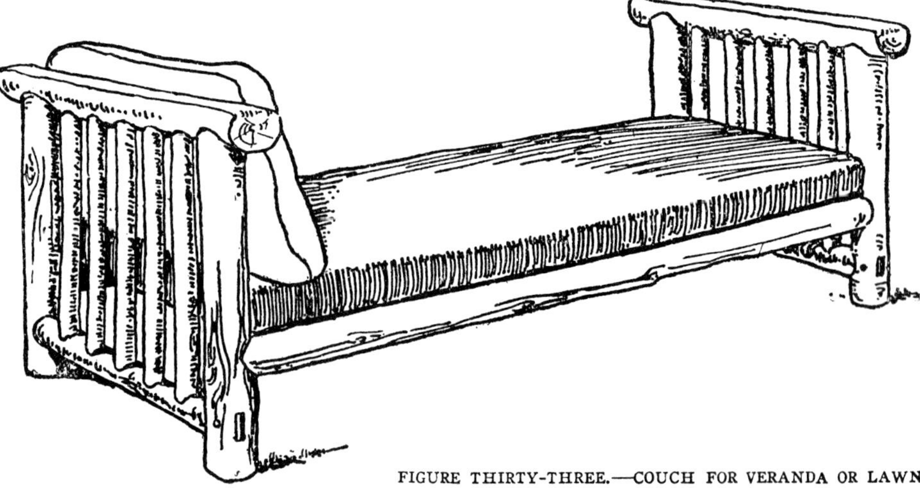

FIGURE THIRTY-THREE.—COUCH FOR VERANDA OR LAWN.

OUR NATIVE WOODS AND THE CRAFTSMAN METHOD OF FINISHING THEM

SO much of the success of the whole Craftsman scheme of building and decoration depends upon the right selection and treatment of the woodwork, which forms such an important part of the structural and also of the decorative scheme, that we have considered it worth while to devote an entire chapter to such information and instruction as we are able to give concerning some of our native woods that we consider most desirable for this purpose. We are taking up only the woods that are native to this country, for the reason that they are nearest at hand and because, when finished by our method, they reveal the beauty of color and grain that forms the basis of the whole Craftsman idea of interior decoration. These vary widely, as each wood possesses strongly marked characteristics as to color, texture and grain; but all the woods we mention here are desirable for interior trim and the use of them is much more in accordance with the Craftsman scheme of decoration than are the elaborate and more or less exotic effects obtained by the use of expensive foreign woods. This does not mean that we claim greater beauty for the native woods, but merely that, when properly treated, they are quite as interesting as any of the more costly woods imported from other countries and have the great advantage of being easily obtainable at moderate cost.

We need not dwell upon the importance of using a generous amount of woodwork to give an effect of permanence, homelikeness and rich warm color in a room. Anyone who has ever entered a house in which the friendly natural wood is used in the form of wainscoting, beams and structural features of all kinds, has only to contrast the impression given by such an interior with that which we receive when we go into the average house, where the plain walls are covered with plaster and paper and the conventional door and window frames are of painted or varnished wood, in order to realize the difference made by giving to the woodwork its full value in the decorative scheme. No care bestowed on decoration, or expense lavished on draperies or furniture, can make up for the absence of wood in the interior of a house. This is a truth that has long been understood and applied in the older countries, especially in England, whose mellow friendly old houses are the delight and despair of Americans; but it is only a few years since we began to apply it to the building and furnishing of our own homes. With us the realization of the possibilities of natural wood when used as a basis for interior decoration first took root in the West, particularly on the Pacific Coast, where the delightful atmosphere of rooms that were wainscoted, ceiled and beamed with California redwood gave rise to a new departure in the finishing and decoration of our homes, and stirred the East to follow suit.

In recommending the generous use of woodwork, however, we would have it clearly understood that we mean the use of wood so finished that its individual qualities of grain, texture and color are preserved so far as possible, and such treatment of wall spaces and structural features that they are not made unduly prominent, but rather sink quietly into the background and become a part of the room itself, forming a friendly unobtrusive setting for the furniture, draperies and ornaments, instead of coming into competition with them. To this end the woodwork should be so finished that its inherent color quality is deepened and mellowed as if by time and its surface made pleasantly smooth without sacrificing the woody quality that comes from frankly revealing its natural texture. When this is done, the little sparkling irregularity of the grain allows a play of light over the surface that seems to give it almost a soft radiance,—a quality that we lose entirely in woodwork that is filled, stained to a solid color, varnished and polished so that the light is reflected from a hard unsympathetic surface.

It is interesting also to note how much the character of a room depends upon the kind of wood we use in it. For example, the impression given by oak is strong, austere and dignified, suggesting stability and permanence such as would naturally belong to a house built to last for generations. It is a robust, manly sort of wood and is most at home in large rooms which are meant for constant use, such as the living room, reception hall, library or dining room. Chestnut, ash and elm,—although each one has an individual quality of color and grain that differentiates it from all the others,—all come into the same class as oak, in that they are strong-fibered, open-textured woods that find their best use in the rooms in which the general life of the household is carried on. The finer-textured woods, such as maple, beech, birch and gumwood, are more suitable for the woodwork in smaller and more daintily furnished rooms that are not so roughly used, such as bedrooms or small private sitting rooms. Aside from this general classification, the choice of wood for interior woodwork naturally must depend upon the taste of the home-builder, the requirements of the decorative scheme planned for the house as a whole, and the ease with which a particular kind of wood may be obtained.

In considering the relative value of our native woods for interior woodwork, we are inclined to give first place to the American white oak, which possesses not only strength of fiber and beauty of color and markings, but great durability, as its sturdiness and the hardness of its texture enables it to withstand almost any amount of wear. In this respect it is far superior to the other woods, such as chestnut, ash and elm, which we have mentioned as being in the same general class of open-textured, strong-fibered woods; although these, under the right treatment, possess a color quality finer than that of oak, in that they show a greater degree of that mellow radiance which counts so much in the atmosphere of a room. This is especially true of chestnut, which is so rich in color that it fairly glows. But in addition to its dignity and durability, there is something about oak that stirs the imagination. Not only is it suggestive of the rich somber time-mellowed rooms of old English houses which have seen generation after generation live and die in them, but it is the wood we are accustomed to associate with nearly all the magnificent carved work of earlier days. In fact, oak has come to stand as a symbol of strength and permanence, and a great part of our affection for it comes from the romance and the rare old associations with which its very name is surrounded.

There are many varieties of oak in this country, but of these the white oak is by far the most desirable, both for cabinetmaking and for interior woodwork. One reason for this is the deep, ripened color it takes on under the process we use for finishing it,—a process which gives the appearance of age and mellowness without in any way altering the character of the wood. We refer to the fuming with ammonia, which we have already described in the preceding chapter. The fact that ammonia fumes will darken new oak was discovered by accident. Some oak boards stored in a stable

in England were found after a time to have taken on a beautiful mellow brown tone and on investigation this change in color was discovered to be due to the ammonia fumes that naturally are present in stables. This ripening, so essential to the beauty of oak woodwork, takes a long time when left to the unaided action of air and sunlight, and the fact that the wood darkened very quickly when it was stored in a stable led to experimenting with the effect of ammonia fumes upon various kinds of oak. The reason for this effect was at first unknown and, to the best of our belief, it was not discovered until the experiments with fuming made in The Craftsman Workshops established the fact that the darkening of the wood was due to the chemical affinity existing between ammonia and tannic acid, of which there is a large percentage present in white oak. This being established, preparations were at once made for using ammonia fumes in a practical way, which we have already described in a preceding chapter. The process mentioned there, however, is practicable only when furniture is to be fumed, as it is quite possible to construct an air-tight compartment sufficiently large to hold one or more pieces of furniture, but when it comes to fuming the woodwork of a whole room it is not so easy. The fuming boxes we use in The Craftsman Workshops are made of tarred canvas stretched tightly over large light wooden frames which are padded heavily around the bottom so that no air can creep in between the box and the floor. The box is drawn to the ceiling by means of a rope and pulley; the furniture is piled directly below and shallow dishes are set around the edges inside the line that marks the limits of the compartment. The box is then lowered almost to the floor; very strong aqua ammonia (26 per cent.) is quickly poured into the dishes and the box dropped at once to the floor. The strength of the ammonia used for this purpose may be appreciated when one remembers that the ordinary ammonia retailed for household use is about 5 per cent.

Of course, for fuming interior woodwork, the air-tight compartment is hardly practicable; but a fairly good substitute for it may be obtained by shutting up the room in which the woodwork is to be fumed, stuffing up all the crevices as if for fumigating with sulphur and then setting around on the floor a liberal number of dishes into which the ammonia is poured last of all. It is hardly necessary to say that the person to whom the pouring of the ammonia is entrusted will get out of the room as quickly as possible after the fumes are released.

Another way of treating oak with ammonia is to brush the liquid directly on the wood, but owing to the strength of the fumes this is not a very comfortable process for the worker and it is rather less satisfactory in its results. The ammonia being in the nature of water, it naturally raises the grain of the wood. Therefore, after the application, it should be allowed to dry over night and the grain carefully sandpapered down the next day. As this is apt to leave the color somewhat uneven, the wood should again be brushed over with the ammonia and sandpapered a second time after it is thoroughly dry. This method of getting rid of the grain is by no means undesirable, for the wood has a much more beautiful surface after all the loose grain has been raised and then sandpapered off. Where paint or varnish is used there is no necessity for getting rid of the grain, as it is held down by them. But with our finish, which leaves the wood very nearly in its natural state, it is best to dispose of the loose grain once for all and obtain a natural surface that will remain permanently smooth.

We find the finest white oak in the Middle West and Southwest, especially in Indiana, which has furnished large quantities of the best grade of this valuable wood. Like so many of our natural resources, the once bountiful supply of our white oak has been so depleted by reckless use that it is probable that ten or fifteen years more will see the end of quartered oak, and possibly of the best grades of plain-sawn oak as well. The popularity of quarter-sawn oak,—a very wasteful process of manufacture,—is one of the causes of the rapid depletion of our oak forests. We append a small cut showing the cross-section of a tree trunk marked with the lines made by quarter-sawing. As will be seen, the trunk is first cut into quarters and then each quarter is sawn diagonally from the outside to the center, naturally making the boards narrower and increasing the waste. There is some hope to be derived from the fact that great stretches of oak timberland are now being reforested by the Government, but at best it will be a generation or two before these slow-growing trees are large enough to furnish the best

quality of lumber. There is no question as to the greater durability of quarter-sawn oak for uses which demand hard wear and also where the finer effects are desired, as in furniture, but for interior woodwork plain-sawn oak is not only much less expensive than quarter-sawn but is quite as desirable in every way. The markings are stronger and more interesting, the difference between the hard and soft parts of the grain is better defined, and the openness of texture gives the wood a mellower color quality than it has when quarter-sawn. The distinguishing characteristic of quarter-sawn oak is the presence of the glassy rays,— technically called medullary rays,—which bind the perpendicular fibers together and give the oak tree its amazing strength. In quarter-sawing, the cut is made parallel with these

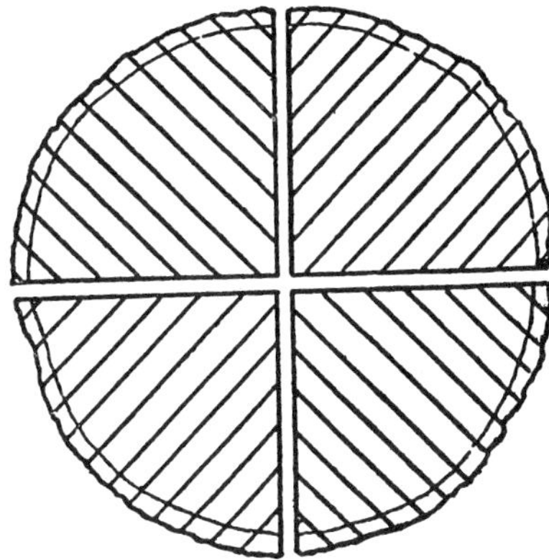

Cross-section of tree-trunk, showing method of quarter-sawing

medullary rays instead of across them, as is done in straight sawing, so that they show prominently, forming the peculiar wavy lines that distinguish quarter-sawn oak. The preservation of the binding properties of these rays gives remarkable structural strength to the wood, which is much less liable to crack, check or warp than when it is plain-sawn. This, of course, makes a difference when it comes to making large panels, table tops, or anything else that shows a large plain surface, and for these uses quarter-sawn oak is preferable merely because it "stands" better. But for the woodwork of a room, we much prefer the plain-sawn oak on account of its friendliness and the delightful play of light and shade that is given by the boldness and color variation of the grain. When quarter-sawn oak is used for large stretches of woodwork, the effect is duller and more austere because the color of the wood is colder and more uniform and it shows a much harder and closer texture.

In the final finishing of oak woodwork, the method that we find most practicable differs somewhat from that described in the directions we have already given for finishing furniture. As the woodwork in a room is not called upon to stand the hard wear that is necessarily given to the furniture, we do not need the shellac, and after the right tone has been obtained by fuming, the wood may be given several coats of prepared floor wax and then rubbed until the surface is satin smooth. If, however, a darker shade of brown is desired, the fumed wood may be given one or more coats of thin shellac, with a little color carried on in each coat, and then finished with wax after the manner described in the directions given for finishing furniture. This method of finishing is one that we have adopted after years of experimenting and it has become so identified with the Craftsman use of oak that it has been very generally taken up by other makers of this style of furniture and by decorators who advocate the Craftsman treatment of interior woodwork.

Next in rank to oak for use in large rooms comes chestnut, which is equally attractive in fiber and markings, has a color quality that is even better, and is plentiful, easily obtained and very reasonable as to cost. While it lacks something of the stateliness and durability of oak, chestnut is even more friendly because of the mellowness and richness of its color, which under very simple treatment takes on a luminous quality that seems to fill the whole room with a soft glow like that of the misty color that is radiated from trees in autumn. Chestnut takes even more kindly than oak to the fuming process, because it contains a greater percentage of tannin and the texture of the wood itself is softer and more open. But unless a deep tone of brown is desired, fuming may be dispensed with, because the wood is so much richer in the elements from which color can be produced that a delightful effect may be obtained merely by applying a light stain of nut brown or soft gray, under which the natural color of the wood appears as an undertone. The staining is very easy to do, but care should be taken to have only a very little color in each coat because the wood takes the stain so readily that a mere trifle of superfluous color will give a thick muddy effect that destroys the clear luminous quality which is

the chief charm. In the case of our Craftsman houses, we find it easier to fume chestnut woodwork than to stain it, and this process is the more to be recommended because chestnut takes the fumes of ammonia very quickly and easily. Also because of this, the ammonia should never be brushed directly on the wood, which is so porous that the moisture is sure to raise the grain to such an extent that the amount of sanding required to smooth it down again destroys the natural surface. One great advantage of chestnut,—aside from its charm of color, texture and markings,—is that it is very easy to work, stays in place readily and is so easy to dry that the chances of getting thoroughly dry lumber are much greater than they would be if oak were used.

Next to chestnut, in our opinion, comes rock elm,—a wood that is fairly abundant, not expensive, and easily obtainable, especially in the East. Rock elm is not affected by the fumes of ammonia and, so far as our experiments go, we have never been able to obtain the right color effect by the use of chemicals. Therefore, in order to get a good color, this wood has to be stained. The colors which are most in harmony with its natural color are brown, green, and gray, particularly in the lighter shades. The distinguishing peculiarity of rock elm is its jagged or feathery grain. Also, the difference in color between the hard and softs parts of the wood is very marked, giving, under the right treatment, a charming variation of tone. If one has the patience to experiment with stains on small pieces of rock elm, some unexpectedly good effects may be obtained. Care must be taken, however, that the stain is light enough to show merely as an over-tone that modifies the natural color of the wood, as the interplay of colors in the grain is hidden by too strong a surface tone. Elm is excellent for interior woodwork where the color effect desired is lighter than that given by either oak or chestnut and also it is hard enough to make pretty good furniture. This last is a decided advantage, especially in a room containing many built-in pieces which naturally form a part of the woodwork. In the earlier days of our experimenting with Craftsman furniture we made a good many pieces of elm and found them, on the whole, very satisfactory.

Brown ash comes into the same class with rock elm, as it is good for furniture as well as interior woodwork. It has a texture and color very similar to elm and should be treated in the same way with a very light stain of either brown, gray or green, all of which blend perfectly with the color quality inherent in the wood. Unfortunately, however, brown ash is no longer plentiful, having been wasted in the same reckless way that we have wasted other excellent woods. Some years ago it was used in immense quantities for making cheap furniture, agricultural implements and the like, and as it was used not only freely but wastefully, the supply is today very nearly exhausted.

In considering all these woods in connection with interior woodwork, it is well to keep in mind that each one of them harmonizes admirably with all the others while retaining, to the full, its own individuality. Therefore, in finishing the rooms on the first floor of a house, it is merely a matter of personal choice as to whether or not the same wood should be used throughout, or each room finished in a different wood. We have often recommended that one wood be used because in a Craftsman house there are practically no divisions or partitions between the rooms, and in this case the effect is so much like that of one large room with many nooks and corners that it would seem the natural thing to use one kind of wood for the interior woodwork throughout. However, if a variation should be desired,— and especially if the separation between the rooms were a little more clearly defined,—the use in different rooms of the different woods we have mentioned would be most interesting, as by this means variety in the woodwork could be obtained without any loss of harmony.

In buildings where it seems desirable to show in the woodwork the bold, strikingly artistic effects such as we associate with Japanese woods, we can heartily recommend cypress, which is plentiful, easily obtained and not expensive. For bungalows, mountain camps, seaside cottages, country clubs and the like, where strong and somewhat unusual effects are sought for, cypress will be found eminently satisfactory, as it is strong and brilliant as to markings and possesses most interesting possibilities in the way of color. Cypress is a soft wood belonging to the pine family and we get most of it from the cypress swamps in the Southern States. It is very like the famous Japanese cypress, which gives such a wonderful charm to many of the Japanese buildings and which is so identified with the Japanese use of woods. Over there they bury

it for a time in order to get the color quality that is most desired,—a soft gray-brown against which the markings stand out strongly and show varying tones. This method, however, did not seem expedient in connection with our own use of the wood and after long experimenting we discovered that we could get much the same effect by treating it with sulphuric acid.

This process is very simple, as it is merely the application of diluted sulphuric acid directly to the surface of the wood. The commercial sulphuric acid should be used rather than the chemically pure, as the first is much cheaper and is quite as good for this purpose. Generally speaking, the acid should be reduced with water in the proportion of one part of acid to five parts of water, but the amount of dilution depends largely upon the temperature in which the work is done. Conditions are best when the thermometer registers seventy-five degrees or more. If it is above that, the sulphuric acid will stand considerably more dilution than it will take if the air is cooler. Of course, in the case of interior woodwork, it is possible to keep the room at exactly the right temperature by means of artificial heat, but when exterior woodwork or shingles are given the sulphuric acid treatment, it is most important to take into consideration the temperature and state of the weather. Exposure to the direct rays of the sun darkens the wood so swiftly that a much weaker solution is required than when the work is done in the shade. In any case, it is best to do a good deal of experimenting upon small pieces of wood before attempting to put the acid on the woodwork itself, as it is only by this means that the exact degree of strength required to produce the best effect can be determined. After the application of the acid the wood should be allowed to dry perfectly before putting on the final finish. For interior woodwork this last finish is given by applying one or two coats of wax; for the exterior, one or two coats of raw linseed oil may be used. If the wood threatens to become too dark under the action of the acid, the burning process can be stopped instantly by an application of either oil or wax, so that the degree of corrosion is largely under the control of the worker. A white hog's-bristle brush should be used for applying the acid, as any other kind of brush would be eaten up within a short time. Also great care should be taken to avoid getting acid on the face, hands, or clothing.

In connection with the subject of cypress for interior woodwork, we desire to say something concerning its desirability for outside use, such as half-timbering and other exterior woodwork. It is one of the most attractive of all our woods for such use because of its color quality and markings and it has the further advantage of "standing" well, without either shrinking or swelling. Naturally the sulphuric acid treatment that we have just described applies to this wood whether it is used indoors or out.

Another use of cypress is found in the rived cypress shingle which give us some of the most interesting effects in exterior wall surfaces. These shingles are the product of one of our few remaining handicrafts, and our sole source of supply depends upon the negroes in the Southern swamps. These negroes are adepts at splitting or riving shingles, and when they get the time or need a little extra money, they split up a few cypress logs into shingles and carry them to a lumber merchant in the nearest town. Consequently, the quantity that is available in the market varies, as no merchant has any great or steady supply of rived shingles and has to accumulate them by degrees and store them, in order to be able to fill any large order. Being hand-rived, these shingles cost about twice as much as the machine-sawn shingles, but they are well worth the extra outlay if one desires a house that is beautiful, individual and durable. The sawn shingle, unless oiled or stained in the beginning, is apt to get a dingy, weather-beaten look under the action of sun and rain and to require renewing early and often. But the rived shingle has exactly the surface of the growing tree from which the bark has been stripped; or, to be more exact, it shows the split surface of a tree trunk from which a bough has been torn, leaving the wood exposed. This surface, while full of irregularities, preserves the smooth natural fiber of the tree, and this takes on a beautiful color quality under the action of the weather, as the color of the wood ripens and shows as an undertone below the smooth silvery sheen of the surface,—an effect which is entirely lost when this natural glint is covered with the "fuzz" left by the saw. These rived shingles are also made of juniper, which is as good in color as cypress and has proven itself even more durable.

All cypress woodwork, whether interior or exterior, takes stain well; and if staining is preferred to the sulphuric acid treatment, very good effects may be gained in this way. We wish, however, to repeat the caution against using too strong a stain, as the effect is always much better if a very little color is carried on in each coat. We cannot too strongly urge the necessity of preliminary experimenting with small pieces of wood in order to gain the best color effects, and we also recommend that in finishing the woodwork of the room itself a very light color be put on at first, to be darkened if a deeper color is found necessary to give the desired effect. The reason for this is that a color which may be considered perfect upon a small piece of wood that is examined closely and held to the light, may prove either too strong or too weak when it is seen on the woodwork as a whole. Much of the effect depends upon the lighting of the room, and therefore it is best to go slowly and "work up" the finish of the woodwork until exactly the right effect is gained. After staining cypress woodwork it should be given either a coat of shellac or wax, or of wax alone, if the amount of wear does not necessitate shellac.

California redwood, when used for interior woodwork, gives an effect as interesting as that obtained by the use of cypress; but redwood does not respond well to the sulphuric acid treatment, which darkens and destroys its beautiful cool pinkish tone. In fact, redwood is best when left in its natural state and rubbed down with wax, as it then keeps in its purity the color quality that naturally belongs to it. Except for this slight finish and protection to the surface, it is a good wood to let alone, as either oil or varnish gives it a hot red look that is disquieting to live with and does not harmonize with any cool tones in the furniture; stains disguise the charm of its natural color and the chemical treatment brings out a purplish tone and gives a darkened and rather muddy effect.

While hard pine is fairly plentiful and lends itself well either to the sulphuric acid treatment or to simple staining, we do not recommend it for interior woodwork, as it costs no less than other woods we have mentioned and is less interesting in color and grain. But if it should be preferred, we would recommend that it be treated with the sulphuric acid, which gives a soft gray tone to the softer parts of the wood and a good deal of brilliancy to the markings.

In considering the woods that are most desirable for woodwork in rooms where light colors and dainty furnishings are used, birch comes first on the list, as it is nearest in character to the open-textured woods we have just described. Of the several varieties, red birch is best for interior woodwork. It is easily obtained all over the East, the Middle West and the South and costs considerably less than the other woods we have mentioned. When left in its natural state and treated with sulphuric acid, red birch makes really beautiful interior woodwork, as the acid deepens its natural color and gives it a mellowness that is as fine in its way as the mellowness produced in oak or chestnut by fuming. Some such treatment is absolutely necessary, for if red birch is left in its natural state, its color fades instead of ripening, so that it gets more and more of a washed-out look as time goes on. In using the acid on birch it is necessary to have a stronger solution than is required in the case of cypress; one part of acid to three parts of water should give it about the required strength. One advantage of birch is its hardness, for after the acid treatment it needs only waxing and rubbing to give it the final finish. The good qualities of birch, treated in this way and used for interior woodwork, are very little known, because it is the wood which has been used more than any other to imitate mahogany. The grain of birch is very similar to that of the more expensive wood, and when it has been given a red water stain and finished with shellac and varnish it bears a close resemblance to mahogany finished in the modern way,— which is by no means to be confused with the rare old Spanish mahogany of the eighteenth century.

Another excellent wood for use in a room that should have comparatively fine and delicate woodwork is maple, which can either be left in its natural color or finished in a tone of clear silver gray. As is well known, the natural maple takes on with use and wear a tone of clear pale yellow. This is not considered generally desirable, but if it should be needed to complete some special color scheme, it can be given to new maple by the careful use of aqua fortis, which should be diluted with water and used like sulphuric acid. The same precautions should be observed in using it, as it is a strong corrosive. Maple is generally

considered much more beautiful when finished in the gray tone, as this harmonizes admirably with the colors most often used in a daintily furnished room,—such as dull blue, old rose, pale straw color, reseda green and old ivory. It is not at all difficult to obtain this gray finish, for all that is needed is to brush a weak solution of iron rust on the wood. This solution is not made by using oxide of iron,— which is commonly but erroneously supposed to mean the same thing as iron rust,—but is obtained by throwing iron filings, rusty nails or any small pieces of iron into acid vinegar or a weak solution of acetic acid. After a couple of days the solution should be strained off and diluted with water until it is of the strength needed to get the desired color upon the wood. It is absolutely necessary in the case of this treatment to experiment first with small pieces of wood before the solution is applied to the woodwork as a whole, because otherwise it would be impossible to judge as to the strength of solution needed to give the desired effect. The color does not show at all until the application is thoroughly dry. If it is too weak, the wood will not be gray enough, and if it is too strong, it will be dark and muddy looking, sometimes almost black. After the woodwork so treated is perfectly dry and has been carefully sandpapered with very fine sandpaper, it should be given a coat of thin shellac that has been slightly darkened by putting in a few drops of black aniline (the kind that is soluble in alcohol); then it is given the final finish by rubbing with wax. These are the only methods we know that give good results on maple. We have tried the sulphuric acid treatment upon this wood, but have not found it satisfactory.

Beech, which is a little darker than maple and of a similar texture and grain, is equally desirable for the same uses. It may be treated either with iron rust or aqua fortis, following the same directions given in the case of maple. This wood is cheap and abundant and is usually found in the same regions which produce birch and maple. Poplar also does very well for the woodwork in a room that is not subjected to hard wear, as it is a very soft wood and will not stand hard usage. The best finish is simply a brown or green stain thin enough to allow the natural color of the wood to show through it. This natural color has in it a strong suggestion of green, so that it affiliates with the green stain and modifies the brown.

One wood that hitherto has been very little known, but that is coming more and more into prominence for the finer sorts of interior woodwork, is gumwood, which is obtained from the red gum that grows so abundantly in the Southern States and on the Pacific Coast. It is a pity that this beautiful wood should have been so little used that most people are unfamiliar with it, because for woodwork where fine texture, smooth surface and delicate coloring are required, quarter-sawn gumwood stands unsurpassed among our native woods. The best effects are obtained from gumwood by treating it with the iron-rust solution used in the way already described in connection with maple; but much more diluted, as the color of gumwood needs only the slightest possible mellowing and toning to make it perfect. When treated with a very weak iron-rust solution it bears a close resemblance to Circassian walnut, and the surface, which is smooth and lustrous as satin, shows a delightful play of light and shade. Sulphuric acid may be used on gumwood, but should be much more diluted than for any other wood, the proportion of acid being not more than one part to eight parts of water. This treatment gives a pinkish cast to the natural gray-brown tone of the wood, and while this does not harmonize as readily with most colors as does the pure gray-brown, it is very effective with certain decorative schemes.

Other woods that are valuable for interior woodwork, although much less plentiful than those we have named, are black walnut, butternut, quartered sycamore and several other woods that come naturally into the same class. Our American black walnut, although one of the standard woods in Europe, has been in a great measure spoiled for us because of its abuse during what we now speak of as the "black walnut period," which has come to mean over-ornamentation, distorted shapes and general bad taste. We have no forests of black walnut left, but there are still single trees, so that if this wood is especially desired, it may be obtained without much difficulty. The characteristics of butternut are much the same as those of black walnut, but it is rather lighter in color and not so hard.

Many people prefer white enameled woodwork for daintily furnished rooms. When this is used, the best kinds of wood for the purpose are poplar and basswood, preferably poplar. One thing should be remembered in connection with white woodwork, and that is that it should be treated in an entirely different way from the typical Craftsman woodwork, which depends for its effect upon the beauty of color and grain and therefore emphasizes these by means of simple forms, straight lines and plain surfaces. When white enameled woodwork is used, the style of it should be more elaborate, as all the interest that naturally belongs to the wood is hidden, and the only way to obtain the play of light and shade necessary to break up the monotony of the white surface is to use moldings, beadings and similar ornamentation, after what is called the Adam style, which we find in the best of our Colonial houses.

In considering interior woodwork one point should not be forgotten; that is the great interest that may be obtained by the right use of what, from a commercial point of view, is faulty wood. We all know the interest and charm of paneling and other woodwork that displays irregularities in the grain, such as knots, knurls and all sorts of queer twists. One of the best examples is found in the "curly" redwood, which is so greatly sought after in California. While the use of such pieces adds greatly to the beauty of a room, the selection of them requires much taste and judgment and absolutely demands that the personal attention of the owner or decorator be given to the work. It is never safe to trust the selection of faulty wood to the lumber merchant or its placing to the carpenter. The necessity of this care is rather an advantage than otherwise, because it is upon just such touches as these that much of the individuality of a decorative scheme depends.

We have treated fully the selection and coloring of the wood, but one practical detail that should be remembered by all who desire beautiful woodwork is that particular attention should be paid to having all the wood thoroughly kiln-dried. Even more important is the necessity of having the house free from dampness before the woodwork is put in, because no wood, however dry and well seasoned, will stand against the dampness of a newly plastered house. In fact, the effect upon the woodwork in such a case is almost worse than when the wood itself is not thoroughly seasoned, for in the latter case it will merely shrink, while dampness in the house will cause it to swell and bulge. The drying of wood not only needs close attention but the aid of some experienced person, as kiln-dried lumber is very apt to be uneven, and there is need of very careful watching while the wood is in the kiln to insure the even drying of all the boards, or the woodwork will be ruined.

Another thing that is worth watching is the final smoothing of the wood before it is put into place. After it leaves the planing machines in the mill it has to be made still smoother, and so most mills that furnish interior trim have installed sandpapering machines. These are convenient and labor-saving, but give a result that is very undesirable for fine woodwork, as the rotary sanding "fuzzes" the grain and, under the light finish we use, it is apt to be raised and roughened by moisture absorbed from the atmosphere. This does not matter when the woodwork is varnished, because the varnish holds it down, but where the natural surface of the wood is preserved great care should be used in the treatment of the grain. The popularity of Craftsman furniture and interior woodwork has created a demand for a surface that shows the sheen of the knife rather than the fuzz of the sanding machine, and some mills have met this demand by putting in scraping machines. These give better results than the sanding machines, but nothing equals the surface that is obtained by smoothing the wood by hand just before it is put into place. For this we use the hand scraper and a smoothing plane that is kept very sharp, as by this method the fiber is cut clean instead of being "cottoned out" and the sheen that naturally belongs to the wood is unimpaired. Although this means hand work, it is not very expensive because of the inconsiderable quantity of wood that is used in a house. Also the Craftsman method of finishing afterward costs so little that the slight extra care and expense incurred in obaining just the right surface is well worth while.

In connection with the woodwork in a house it is necessary to give some attention to the floors, which come into close relation with the treatment of the walls. The best wood for flooring is quartered oak, which all lumber merchants keep in stock in narrow widths, tongued and grooved. We find, however, that a more interesting floor can be made by using wider boards of uneven width, as this gives an effect of strength and bigness to the room. These wide boards need not be tongued and grooved, but may be put together with butt joints and the boards nailed through the top by using brad-head nails that can be countersunk and the holes puttied up so that they are almost invisible. When very wide boards are used it is best to build the floor in "three ply," like paneling. Plain-sawn oak is also good for flooring, but it is more likely to warp and sliver than quartered oak and it does not lie so flat. An oak floor, whether plain or quarter-sawn, must always be filled with a silex wood filler so that its surface is made smooth and non-absorbent. The color should be made the same as that of the woodwork, or a little darker; and after the stain is applied, the floor should be given one coat of shellac and then waxed. In rooms where the color schemes permit a slightly reddish tone in the floor, we would suggest that either birch or beech be used for flooring, as these may be finished by the sulphuric acid process,—a method which is better than stain because it darkens the wood itself and therefore does not wear off with use. If a gray floor should be desired, we would suggest maple treated with the iron-rust solution. In either case a coat of thin shellac should be applied after the chemical has been thoroughly dried,—say twenty-four hours after the first application,— and then waxed in the regular way. For ordinary floors a good wood to use is comb-grained pine, which receives its name from the method of sawing that leaves the grain in straight lines, not unlike the teeth of a comb. This does not warp or sliver and is very durable; it may be treated with stain and then given the regular finish of shellac and wax.

THE CRAFTSMAN FIREPLACE: A COMPLETE HEAT-ING AND VENTILATING SYSTEM

O N account of the number of inquiries that have been received in regard to the Craftsman fireplace-furnace, I have thought it advisable to publish another description and set of drawings, embodying several improvements over the form of fireplace originally shown. These improvements have simplified not only the construction of the heater but also the work of installing it.

As shown in Figures 1, 2 and 3, the heater body is made of large sheets of steel, welded together by special welding machinery into

VIEW OF CRAFTSMAN FIREPLACE OF TAPESTRY BRICK, WITH OPEN HEARTH AND ANDIRONS FOR BURNING WOOD.

one piece of continuous metal, making leakage of gas, smoke or dust an impossibility. It is so constructed that each smoke compartment is self-cleaning; the smoke areas being vertical, there is no place for soot and moisture to collect. The heater is six feet high and four feet wide and weighs complete with grates and other iron parts needed in the construction about 1,000 pounds.

Grates for the burning of coal or coke are supplied with each heater. These consist of only three parts and are easily and quickly set in place. Figure 1 shows the removable metal hearth and grates in place for the burning of coal. The ashes sift through the grate into the ash pit, which is so large that it needs emptying only once a season. This also eliminates the objectionable feature of dust from the ashes escaping into the room. If it is desired to burn wood upon an open hearth, the metal hearth and grates are removed and the opening into the ash pit is covered by a metal plate on which andirons may be placed for burning wood, as shown in the photograph. Whether the fireplace is equipped with hearth and grates for the burning of coal, or is arranged with open hearth and andirons for the burning of wood, the impression given is at once satisfactory and permanent.

The heater is set on the floor level, and the installation consists in merely building a four-inch brick wall around it. This wall, carried up to the ceiling and roofed over, forms the warm air chamber above the furnace body. In one leg of the chimneypiece is set a smoke flue, shown by dotted lines in Figure 2, which is connected with the body of the heater. I furnish one section of this flue lining, having three holes: one which fits onto the flange around the smoke outlet of the heater, another which may be connected with a pipe from the kitchen range, and a third which opens by register into the room for the purpose of checking the coal fire, but which is

VERTICAL SECTION THROUGH CENTER OF CRAFTSMAN FIREPLACE.

kept closed when wood is burned. This flue starts at the bottom of the smoke outlet on the heater, shown in Figure 3, leaving the leg

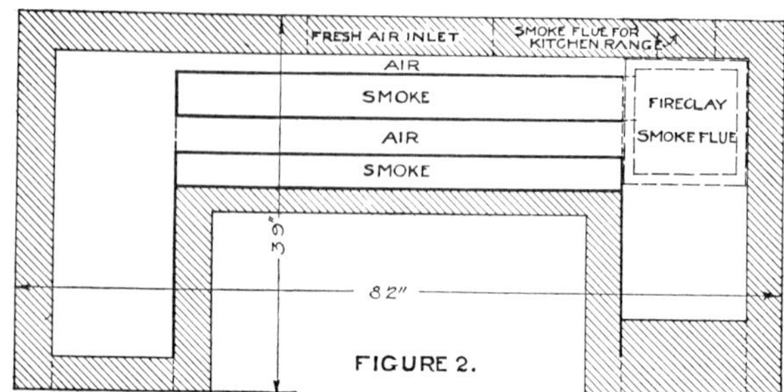

SECTIONAL PLAN OF THE CRAFTSMAN FIREPLACE.

of the chimney below the flue free for the circulation of air.

Any mason can build the wall and make a correct installation. The cost of the brickwork complete with chimney is less than for the usual fireplace of equal size. About 3,000 brick are required where there is a cellar and the chimney is carried up two stories. At a cost of $10.00 per thousand for brick and an equal sum per thousand for sand, cement and labor, the entire cost of brickwork, including $5.00 for flue lining, would be about $65.00.

I have used the common hard-burned brick as a basis for the above figures; where the owner desires to make the fireplace of plaster, stone, Tapestry brick or tile, the additional cost will depend upon the material selected. Hard-burned brick laid up with a wide mortar joint will make a beautiful fireplace. There are no limitations as to the design of the chimneypiece, the only requirement being that the inside measurements are kept to those shown in Figure 2.

The Craftsman fireplace may be installed in houses already built, as well as in new ones, the work in each case requiring a new chimney and the cost being practically the same. Because of the universal favor of the open fire it seems best to make only a medium-sized heater, as many people would prefer to have two or even more fireplaces in different parts of the house. The piping of warm air to the various rooms is then a small factor of the cost, as the pipes will be few and short. These are to be furnished by the owner, since they are always in stock at the local hardware store and are inexpensive.

The operation of the heater itself is as follows: As shown in Figure 1 the smoke generated by the fire passes up through the smoke compartment, down behind the steel smoke wall to the bottom of the heater, then up through another smoke compartment to the smoke outlet flue shown in Figure 3, and out through the chimney.

During its passage the smoke heats the steel walls of the smoke compartments, which in turn heat the air in the air compartments, as will be seen by reference to Figures 1 and 2. The air is thus caused to rise and pass up into the warm air chamber. This action draws in outside air through the fresh air inlet, up through the air compartments into the warm air chamber. At the same time air is also being drawn in from the room through the registers at the base of the fireplace, up through the air compartments into the warm air chamber, where it mixes with the warmed fresh air from outside. The warmed air passes through the upper registers into rooms on the first floor and also through the air pipes to the upper rooms. These air pipes and registers are proportioned in size so that each one will deliver the proper amount of air to the various rooms.

The warm air, upon entering each of the upper rooms, being lightest, rises and spreads out in an even layer against the ceiling. This layer, as it cools, descends to the floor and passes out under the door, down the stairway opening to the lower floor. Part of this air is drawn into the fire and passes out through the chimney, and the rest is drawn into the lower registers. The circulation is rapid and positive, being accomplished, as seen, by gravitation, the heavier or colder air seeking the lowest level and the lighter or warmer air the highest. The heater thus maintains a constant circulation between the various rooms as well as a movement of the air within the rooms, making a given air supply go much farther than with other heating systems.

In this circulation, the air absorbs all impurities, and naturally the zone of the most vitiated air is nearest the floor. It is from this zone that the fireplace draws immense quantities of air and discharges it through the chimney. An adult vitiates from 2,500 to 3,000 cubic feet of air per hour. The fireplace is constructed to admit 20,-000 cubic feet of fresh air and discharge through the chimney the same amount of vitiated or used air per hour, thus making perfect ventilation for seven adults. In this way the air throughout the house is entirely replaced with fresh warmed air from outdoors every fifteen or twenty minutes. Doors and windows should be kept closed in order that the circulation of air may not be disturbed, for upon the proper circulation depends the efficient heating and ventilating of the house. Under these conditions there can be no drafts.

The danger of the fireplace smoking is entirely eliminated, as the smoke and air openings are properly proportioned and, being part of the steel body, do not depend upon the judgment of the mason. Moreover, it is not only impossible for back drafts to force smoke into the room, but sparks are prevented from escaping through the flue, thereby removing all danger of fire on the roof.

The conserving within the brick walls of all heat which has formerly been lost in the cellar; the circulating of volumes of air in contact with the large areas of smoke surface, thereby extracting practically all the heat from the smoke, and the radiation of heat direct into the room from the open fire, make the Craftsman fireplace a most efficient heating system. One fireplace will amply heat a seven-room house, with a consumption of from seven to ten tons of coal per year in a climate like that of our Central States. The exact amount of fuel consumed, however, depends largely upon the exposure, the number and size of the windows, and the construction of the house. Coal or coke will furnish a more even and steady heat both day and night than wood, but because of the slow combustion due to the down draft, wood may be used as a fuel with entire satisfaction from a standpoint of both economy and attention. Then, too, the wood fire is so much quicker than coal or coke in producing heat and so easily started that its use will be almost universal for fuel during the late fall and early spring.

The price of the steel heater complete with grates, registers and all metal parts (except the pipes needed to conduct warm air from the heater to rooms distant from it) amounts, with the freight, to $150.00. By combining this with the cost of the brickwork and the pipes one can easily install the heating plant complete inside of $250.00. The fireplace is sold only direct to users. I require the plans of each house in which it is to be installed, and from them I make and furnish free to the owner a heating layout which shows the location and size of warm air pipes and registers, and includes complete plans and instruc-

tions for the mason to use in building the brickwork. I guarantee the fireplace to heat and ventilate properly each house in which it is installed, and by making the heating plant myself and selling it direct to users, I am in a position to assume the entire responsibility of its giving satisfaction.

I am ready to make shipments of Craftsman fireplaces, and shall be glad to hear from all those who are considering the installation of

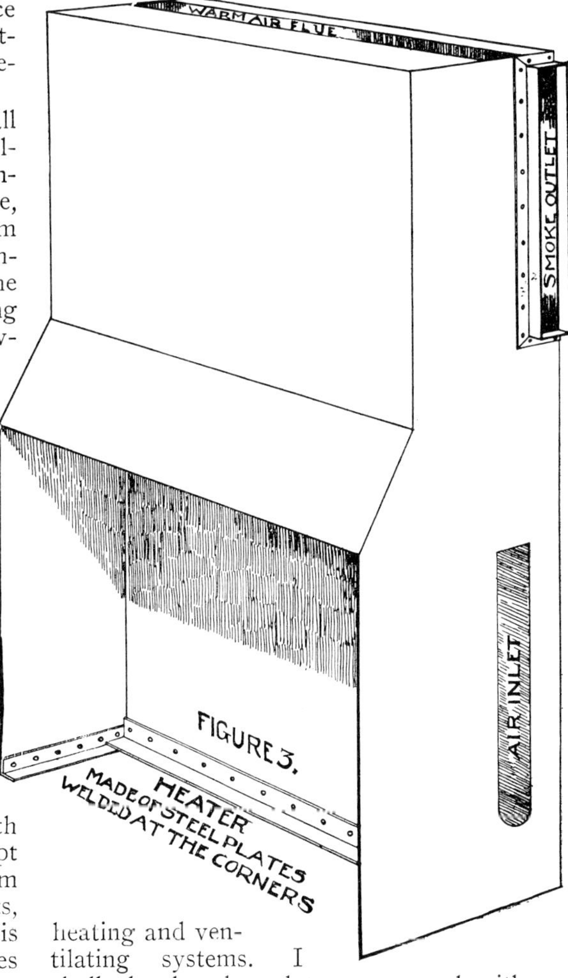

heating and ventilating systems. I shall also be pleased to correspond with any readers who may wish further explanation of the construction of the heater and the manner of its operation.

THE CRAFTSMAN IDEA OF THE KIND OF HOME ENVIRONMENT THAT WOULD RESULT FROM MORE NATURAL STANDARDS OF LIFE AND WORK

IN this book we have endeavored to set forth as fully as possible the several parts which, taken together, go to make up the Craftsman idea of the kind of home environment that tends to result in wholesome living. We have shown the gradual growth of this idea, from the making of the first pieces of Craftsman furniture to the completed house which has in it all the elements of a permanently satisfying home. But we have left until the last the question of the right setting for such a home and the conditions under which the life that is lived in it could form the foundation for the fullest individual and social development.

There is no question now as to the reality of the world-wide movement in the direction of better things. We see everywhere efforts to reform social, political and industrial conditions; the desire to bring about better opportunities for all and to find some way of adjusting economic conditions so that the heart-breaking inequalities of our modern civilized life shall in some measure be done away with. But while we take the greatest interest in all efforts toward reform in any direction, we remain firm in the conviction that the root of all reform lies in the individual and that the life of the individual is shaped mainly by home surroundings and influences and by the kind of education that goes to make real men and women instead of grist for the commercial mill.

That the influence of the home is of the first importance in the shaping of character is a fact too well understood and too generally admitted to be offered here as a new idea. One need only turn to the pages of history to find abundant proof of the unerring action of Nature's law, for without exception the people whose lives are lived simply and wholesomely, in the open, and who have in a high degree the sense of the sacredness of the home, are the people who have made the greatest strides in the development of the race. When luxury enters in and a thousand artificial requirements come to be regarded as real needs, the nation is on the brink of degeneration. So often has the story repeated itself that he who runs may read its deep significance. In our own country, to which has fallen the heritage of all the older civilizations, the course has been swift, for we are yet close to the memory of the primitive pioneer days when the nation was building, and we have still the crudity as well as the vigor of youth. But so rapid and easy has been our development and so great our prosperity that even now we are in some respects very nearly in the same state as the older peoples who have passed the zenith of their power and are beginning to decline. In our own case, however, the saving grace lies in the fact that our taste for luxury and artificiality is not as yet deeply ingrained. We are intensely commercial, fond of all the good things of life, proud of our ability to "get there," and we yield the palm to none in the matter of owning anything that money can buy. But, fortunately, our pioneer days are not ended even now and we still have a goodly number of men and women who are helping to develop the country and make history merely by living simple natural lives close to the soil and full of the interest and pleasure which come from kinship with Nature and the kind of work that calls forth all their resources in the way of self-reliance and the power of initiative. Even in the rush and hurry of life in our busy cities we remember well the quality given to the growing nation by such men and women a generation or two ago and, in spite of the chaotic conditions brought about by our passion for money-getting, extravagance and show, we have still reason to believe that the dominant characteristics of the pioneer yet shape what are the salient qualities in American life.

To preserve these characteristics and to bring back to individual life and work the vigorous constructive spirit which during the last half-century has spent its activities in commercial and industrial expansion, is, in a nut-shell, the Craftsman idea. We need to straighten out our standards and to get rid of a lot of rubbish that we have accumulated along with our wealth and commercial supremacy. It is not that we are too energetic, but that in many ways we have wasted and misused our energy precisely as we have wasted and misused so many of our wonderful natural resources. All we really need is a change in our point of view toward life and a keener perception regarding the things that count and the things which merely burden us. This being the case, it would seem obvious that the place to begin a readjustment is in the home, for it is only natural that the relief from friction which would follow the ordering of our lives along more simple and reasonable lines would not only assure greater comfort, and therefore greater efficiency, to the workers of the nation, but would give the children a chance to grow up under conditions which would be conducive to a higher degree of mental, moral and physical efficiency.

THEREFORE we regard it as at least a step in the direction of bringing about better conditions when we try to plan and build houses which will simplify the work of home life and add to its wholesome joy and comfort. We have already made it plain to our readers that we do not believe in large houses with many rooms elaborately decorated and furnished, for the reason that these seem so essentially an outcome of the artificial conditions that lay such harassing burdens upon modern life and form such a serious menace to our ethical standards. Breeding as it does the spirit of extravagance and of discontent which in the end destroys all the sweetness of home life, the desire for luxury and show not only burdens beyond his strength the man who is ambitious to provide for his wife and children surroundings which are as good as the best, but taxes to the utmost the woman who is trying to keep up the appearances which she believes should belong to her station in life. Worst of all, it starts the children with standards which, in nine cases out of ten, utterly preclude the possibility of their beginning life on their own account in a simple and sensible way. Boys who are brought up in such homes are taught, by the silent influence of their early surroundings, to take it for granted that they must not marry until they are able to keep up an establishment of equal pretensions, and girls also take it as a matter of course that marriage must mean something quite as luxurious as the home of their childhood or it is not a paying investment for their youth and beauty. Everyone who thinks at all deplores the kind of life that marks a man's face with the haggard lines of anxiety and makes him sharp and often unscrupulous in business, with no ambition beyond large profits and a rapid rise in the business world. Also we all realize regretfully the extravagance and uselessness of many of our women and admit that one of the gravest evils of our times is the light touch-and-go attitude toward marriage, which breaks up so many homes and makes the divorce courts in America a by-word to the world. But when we think into it a little more deeply, we have to acknowledge that such conditions are the logical outcome of our standards of living and that these standards are always shaped in the home.

That is why we have from the first planned houses that are based on the big fundamental principles of honesty, simplicity and usefulness,—the kind of houses that children will rejoice all their lives to remember as "home," and that give a sense of peace and comfort to the tired men who go back to them when the day's work is done. Because we believe that the healthiest and happiest life is that which maintains the closest relationship with out-of-doors, we have planned our houses with outdoor living rooms, dining rooms and sleeping rooms, and many windows to let in plenty of air and sunlight. The most cursory examination of the floor plans given in this book will show that we have put into practical effect our conviction that a house, whatever its dimensions, should have plenty of free space unencumbered by unnecessary partitions or over-much furniture. Therefore we have made the general living rooms as large as possible and not too much separated one from the other. It seems to us much more friendly, homelike and comfortable to have one big living room into which one steps directly from the entrance door,—or from a small vestibule if the climate demands such a protection,—and to have this living room the place where all the business and pleasure of the common family life may be carried on. And we like it to have pleasant nooks and corners which give a comfortable sense of semi-privacy and yet are not in any way shut off from the larger life of the room. Such an arrangement has always seemed to us symbolic of the ideal conditions of social life. The big hospitable fireplace is almost a necessity, for the hearthstone is always the center of true home life, and the very spirit of home seems to be lacking when a register or radiator tries ineffectually to take the place of a glowing grate or a crackling leaping fire of logs.

Then too we believe that the staircase, instead of being hidden away in a small hall or treated as a necessary evil, should be made one of the most beautiful and prominent features of the room, because it forms a link between the social part of the house and the upper regions which belong to the inner and individual part of the family life. Equally symbolic is our purpose in making the dining room either almost or wholly a part of the living room, for to us it is a constant expression of the fine spirit of hospitality to have the dining room, in a way, open to all comers. Furthermore, such an arrangement is a strong and subtle influence in the direction of simpler living because entertainment under such conditions naturally grows less elaborate and more friendly,—less alien to the regular life of the family and less a matter of social formality.

Take a house planned in this way, with a big living room made comfortable and homelike and beautiful with its great fireplace, open staircase, casement windows, built-in seats, cupboards, bookcases, sideboard and perhaps French doors opening out upon a porch which links the house with the garden; fill this room with soft rich restful color, based upon the mellow radiance of the wood tones and sparkling into the jeweled high lights given forth by copper, brass, or embroideries; then contrast it in your own mind with a house which is cut up into vestibule, hall, reception room, parlor, library, dining room and den,—each one a separate room, each one overcrowded with furniture, pictures and bric-a-brac,—and judge for yourself whether or not home surroundings have any power to influence the family life and the development of character. If you will examine carefully the houses shown in this book, you will see that they all form varying expressions of the central idea we have just explained, although each one is modified to suit the individual taste and requirements of the owner. This is as it should be, for a house expresses character quite as vividly as does dress and the more intimate personal belongings, and no man or woman can step into a dwelling ready made and decorated according to some other person's tastes and preferences without feeling a sense of strangeness that must be overcome before the house can be called a real home.

It will also be noticed in examining the plans of the Craftsman houses that we have paid particular attention to the convenient arrangement of the kitchen. In these days of difficulties with servants and of inadequate, inexperienced help, more and more women are, perforce, learning to depend upon themselves to keep the household machinery running smoothly. It is good that this should be so, for woman is above all things the home maker and our grandmothers were not far wrong when they taught their daughters that a woman who could not keep house, and do it well, was not making of her life the success that could reasonably be expected of her, nor was she doing her whole duty by her family. The idea that housekeeping means drudgery is partly due to our fussy, artificial, overcrowded way of living and partly to our elaborate houses and to inconvenient arrangements. We believe in having the kitchen small, so that extra steps may be avoided, and fitted with every kind of convenience and comfort; with plenty of shelves and cupboards, open plumbing, the hooded range which carries off all odors of cooking, the refrigerator which can be filled from the outside,—in fact, everything that tends to save time, strength and worry. In these days the cook is an uncertain quantity always and maids come and go like the seasons, so the wise woman keeps herself fully equipped to take up the work of her own house at a moment's notice, by being in such close touch with it all the time that she never lays down the reins of personal government. The Craftsman house is built for this kind of a woman and we claim that it is in itself an incentive to the daughters of the house to take a genuine and pleasurable interest in household work and affairs, so that they in their turn will be fairly equipped as home makers when the time comes for them to take up the more serious duties of life.

WE HAVE set forth the principles that rule the planning of the Craftsman house and have hinted at the kind of life that would naturally result from such an environment. But now comes one of the most important elements of the whole question,—the surroundings of the home. We need hardly say that a house of the kind we have described belongs either in the open country or in a small village or town, where the dwellings do not elbow or crowd one another any more than the people do. We have planned houses for country living because we firmly believe that the country is the only place to live in.

The city is all very well for business, for amusement and some formal entertainment,—in fact for anything and everything that, by its nature, must be carried on outside of the home. But the home itself should be in some place where there is peace and quiet, plenty of room and the chance to establish a sense of intimate relationship with the hills and valleys, trees and brooks and all the things which tend to lessen the strain and worry of modern life by reminding us that after all we are one with Nature.

Also it is a fact that the type of mind which appreciates the value of having the right kind of a home, and recognizes the right of growing children to the most natural and wholesome surroundings, is almost sure to feel the need of life in the open, where all the conditions of daily life may so easily be made sane and constructive instead of artificial and disintegrating. People who think enough about the influence of environment to put interest and care into the planning of a dwelling which shall express all that the word "home" means to them, are usually the people who like to have a personal acquaintance with every animal, tree and flower on the place. They appreciate the interest of planting things and seeing them grow, and enjoy to the fullest the exhilarating anxiety about crops that comes only to the man who planted them and means to use them to the best advantage. Then again, such people feel that half the zest of life would be gone if they were to miss the fulness of joy that each returning spring brings to those who watch eagerly for the new green of the grass and the blossoming of the trees. They feel that no summer resort can offer pleasures equal to that which they find in watching the full flowering of the year; in seeing how their own agricultural experiments turn out, and in triumphing over each success and each addition to the beauty of the place that is their own. Few of these people, too, would care to miss the sense of peace and fulfilment in autumn days, when the waning beauty of the year comes into such close kinship with the mellow ripeness of a well-spent life that has borne full fruit. And what child is there in the world who would spend the winter in the city when there are ice-covered brooks to skate on, the comfort of jolly evenings by the fire and the never-ending wonder of the snow? And all the year round there are the dumb creatures for whom we have no room or time in the city,—the younger brothers of humanity who submit so humbly to man's dominion and look so placidly to him for protection and sustenance.

THANK heaven, though, we are not so far away from our natural environment that it needs much to take us back to it. We have many evidences of the turning of the tide of home life from the city toward the country. Even workers in the city are coming more and more to realize that it is quite possible to maintain their place in the business world and yet give their children a chance to grow up in the country. Also the economic advantage of building a permanent home instead of paying rent year after year is gaining an ever-increasing recognition, so that in a few years the American people may cease to deserve the reproach of being a nation of flat-dwellers and sojourners in family hotels. The instinct for home and for some tie that connects us with the land is stronger than any passing fashion, and although we have in our national life phases of artificiality that are demoralizing they affect only a small percentage of the whole people, and when their day is over they will be forgotten as completely as if they had never existed. Psychologists talk learnedly of "Americanitis" as being almost a national malady, so widespread is our restlessness and feverish activity; but it is safe to predict that, with the growing taste for wholesome country life, it will not be more than a generation or two before our far-famed nervous tension is referred to with wonder as an evidence of past ignorance concerning the most important things of life.

And when we have turned once more to natural living instead of setting up our puny affairs and feverish ambitions to oppose the quiet, irresistible course of Nature's law, we will not need to turn hungrily to books for stories of a bygone Golden Age, nor will we need to deplore the vanishing of art and beauty from our lives, for when the day comes that we have sufficient courage and perception to throw aside the innumerable petty superfluities that hamper us now at every turn and the honesty to realize what Nature holds for all who turn to her with a reverent spirit and an open mind, we will find that art is once more a part of our daily life and that the impulse to do beautiful and vital creative work is as natural as the impulse to breathe.

Therefore it is not idle theorizing to prophesy that, when healthful and natural conditions are restored to our lives, handicrafts will once more become a part of them, because two powerful influences will be working in this direction as they have worked ever since the earliest dawn of civilization. One is the imperative need for self-expression in some form of creative work that always comes when the conditions of life are such as to allow full development and joyous vigor of body and mind. The other is that which closer relationship with Nature seems to bring; a craving for greater intimacy with the things we own and use. Machine-made standards fall away of themselves as we get away from artificial conditions. It is as if wholesome living brought with it not only quickened perceptions but also a sense of personal affection for all the familiar surroundings of our daily life. It is from such feeling that we get the treasured heirlooms which are handed down from generation to generation because of their associations and what they represent.

Naturally the primitive conditions of pioneer life in any nation include handicrafts as a matter of course, from the simple fact that people had to make for themselves what they needed or go without. We realize that in this age of invention and of labor-saving machinery it is neither possible nor desirable to return to such conditions, but we believe that it is quite possible for a higher form of handicrafts to exist under the most advanced modern conditions and that achievements as great as those of the old craftsmen who made famous the Mediæval guilds are by no means out of the reach of modern workers when they once realize the possibilities that lie in this direction. Our theory is that modern improvements and conveniences afford a most welcome and necessary relief from the routine drudgery of household and farm work by disposing quickly and easily of what might much better be done by machinery than by hand, and that therefore there should be sufficient leisure left for the enjoyment of life and for the doing of work that is really worth while, which are among the things most essential to all-round mental and moral development. Almost the greatest drawback to farm life as it is today is the lack of interest and of mental alertness.

Especially is this the case during the winter months, when work on the farm is slack and much time is left to be spent in idleness or in some trifling occupation. Consider what the effect would be if it were made possible at such times to take up some form of creative work that would not only bring into play every atom of interest and ability, but would also serve a practical purpose by adding considerably to the family income!

WE HAVE given a great deal of consideration to the practical side of such a combination of handicrafts and farming, and we realize of course that the great difficulty in the way of making such a thing possible by making it profitable is the question of obtaining a steady market for the products of such crafts as might be practiced in connection with country life. It is often urged as an argument against handicrafts that hand-made goods could not possibly compete with factory-made goods, and that it would be absurd for people to waste time in making things for which there would be no sale. This does not seem to us to be the case, for the reason that there is no competition between the products of handicrafts and factory-made goods, because they are not measured by the same standard of value nor do they appeal to the same class of consumer. Hand-made articles have a certain intrinsic value of their own that sets them entirely apart from machine-made goods. This value depends, not upon the fact that the article is made entirely by hand or with appropriate tools,—that is not the point,—but upon the skill of the workman, his power to appreciate his own work sufficiently to give it the quality that appeals to the cultivated taste and the care that he gives to every detail of workmanship, from the preparation of the raw material to the final finish of the piece. We are not urging that handicrafts be cultivated in connection with farming for the purpose of competing with the factories for the same class of trade, for, with the demand that necessitates the immense production of goods of all kinds, the labor-saving machinery and efficient methods of the factories are absolutely essential, just as they are essential in the general economic scheme because they furnish employment to thousands of workers who ask nothing better than to be allowed to tend a machine with a certainty of so much a day coming to them at the end of the week. The place of home and village industries on the economic side, is to supplement the factories by producing a grade of goods which it is impossible to duplicate by machinery,—and which command a ready market when they can be found,—and to give to the better class of workers a chance not only to develop what individual ability they may possess, but to reap the direct reward of their own energy and industry in the feeling that they are free of the wage system with all its uncertainties and that what they make goes to maintain a home that is their own, to educate their children and to lay up a sufficient provision against old age.

We do not deny that handicrafts, as practiced by individual arts and crafts workers in studios, fall very short of affording a sufficient living to craft workers as a class, and also we do not deny that small farming as carried on in our thinly populated districts is neither interesting, pleasant, nor profitable. But we do assert that it is possible to connect the two and to carry them on upon a basis that will insure not only peace and comfort in living, and a form of industry that affords the greatest opportunity for all-round development, but also a permanent competence. To bring about such a condition is the end and aim of the whole Craftsman idea. We call it by that name because we have been the first to formulate it in this country. But it is in the air everywhere. It is taking shape in several of the European countries in the form of government appropriations for the reëstablishment and encouragement of handicrafts among the people, government schools for the teaching of various crafts, and government exchanges to look after the question of a steady market. In Great Britain and Ireland the same thing is being done by private enterprise, partly as a matter of social reform and partly as an effort of philanthropy. But in this country conditions are different. We have no peasant class and almost the only people in need of social reform, or of philanthropic efforts in their behalf are the vast hordes of immigrants who pour into the country each year and too often find it difficult to adjust their lives to American conditions.

THE people to whom the Craftsman idea makes its appeal are the better class farmers who own their farms, workers in the city who are able to get together a little place in the country and build up a permanent home, and the better class of artisans who desire to escape from the routine of factory work. That such people are taking a keen interest in the question of life in the country and that farming is rapidly being restored to its former status as a desirable occupation is evidenced by the encouragement given to the widespread activities of the Department of Agriculture, which is doing so much to bring about better and more economical methods of cultivating the soil. We have plenty of proof that these efforts do not fall short in the matter of results, for all over the country there is a growing appreciation of the possibilities that lie in intensive agriculture and a desire to learn something of modern scientific farming. We most heartily endorse all that is being done along these lines; but we go a step farther because we maintain that the whole standard of living must be changed before there can be a return of natural conditions to our lives. For example, we have been accustomed of late years to an artificial scale of income and expenditure, and the prices of the most ordinary necessities of life have risen so high that it takes all the average man can do to make ends meet. This is both wrong and unnecessary, but a natural consequence of artificial conditions, and we maintain that the only way to correct it is to put ourselves in a position to realize that, in permitting our lives to be ruled by false standards and inflated values, we have lost sight of the principle that economy means wealth. When we regain this simple and reasonable point of view, we will find no difficulty in admitting that comfort and happiness in living do not depend upon the amount of money we can make and spend, but upon pleasant surroundings and freedom from the pressure of want and apprehension; and when this truth is brought home to the affairs of daily life, the work of establishing natural standards is done.

Therefore we advocate a return to cultivating the soil as a means of obtaining the actual living,—that is, of looking to garden, grain-patch, orchard, chicken yard and pasture for the vegetables, fruits, cereals, eggs and meat consumed by the family. If properly cared for and cultivated according to the modern methods that are now everybody's for the learning, a little farm of five or ten acres can be made not only to yield a living for its owner and his family but a handsome surplus for the markets, thus serving the double purpose of stopping the outflow and adding to the income of actual money as well as providing home comfort and healthful working surroundings. The farm home once established, its owner is free of any steady expense save for taxes and repairs, so that everything that is done is constructive and cumulative in its effects.

This, in brief, is the whole idea of the Craftsman home,—a pleasant comfortable dwelling situated on a piece of ground large enough to yield, under proper cultivation, a great part of the food supply for the family. Such a home, by its very nature, would be permanent and, with the right kind of education and healthful occupation for the children, would do much to stop the flow of population into the great commercial centers and to insure a more even division of prosperity throughout the land. In many instances the home is an established fact, but the education and the occupation are yet to come. It is with a view to solving this problem that we advocate individual handicrafts in the home and industries to be carried on upon a more extended scale in the neighborhood or the village. The very fact of a thorough training in any useful craft would insure to a boy or girl the right groundwork for an education, so that the solution of one problem is practically a solution of both.

Naturally, the greatest field for home handicrafts lies in the making of household furnishings, wearing apparel and articles of daily use. For example, there is a large and steady demand for hand-woven, hooked and hand-tufted rugs in good designs and harmonious colorings, especially when they can be had at reasonable prices. That there would be a market for good hand-made rugs in this country is shown by the demand for similar rugs that are made abroad by peasant labor. This is, of course, much cheaper than any class of labor in this country. Nevertheless, the same grade of rugs could be made here by home and farm workers and sold at a profit at the same price that must be demanded for the imported rugs, after the high import duty on this class of goods has been added to the original cost. Also cabinetmaking, considered as a handicraft, opens a field of unusually wide and varied interest, as the making of things so closely associated with our daily life and surroundings is a form of work that is both delightful and profitable. Iron work is equally interesting, and a preliminary training in good hard blacksmithing not only offers an excellent foundation for the doing of good things in structural iron work and articles for household use, but it equals wood work in developing any creative power that may be latent in the worker. Weaving and needlework come into the first rank of interesting and profitable crafts, and among the lighter industries that offer a chance for individual expression and at the same time pay pretty well, are basketry, block printing, dyeing, lace making, bookbinding and the like.

EVER since its first publication in nineteen hundred and one, THE CRAFTSMAN Magazine has, in one form or another, been advocating this idea; and we have most satisfactory proof in the growth and standing of the magazine that a great many people in this country are thinking along the same lines. When THE CRAFTSMAN was founded it was with the intention of making it a magazine devoted almost solely to the encouragement of handicrafts in this country.   We believed, then, as we believe now, in the immense influence for good in the development of character that is exerted merely by learning to use the hands.   One needs only to look at any part of the history of handicrafts to realize how much strength, sincerity and genuine creative thought went into the work of the old craftsmen who were also such solid and substantial citizens. We have always felt that it is not the making of things that is important, but the making of strong men and women through the agency of the sound development that begins when the child learns to use its hands for shaping to the best of its ability something which is really needed either for its own play or for the comfort and convenience of others in the home.   It is going back in spirit to the primitive beginning of handicrafts,—which marks the beginning of civilization, and is so important a factor in the growth of character that upon it depends nearly every quality of heart and brain that goes into what we may call the craftsmanship of life.   But, as THE CRAFTSMAN grew and step by step attained a wider outlook, the question of the study of handicrafts as an end in itself gradually sunk to a position of minor importance in the policy of the magazine.   Our belief that in it lay the foundation of all growth was no less, but the field was so broad that the record and discussion of all constructive work in the larger affairs of life came gradually to take first place.

As we began to design houses and to shape the idea of the Craftsman country home as we have here tried to describe it, we took up the subjects of architecture and interior decoration, doing our best to promote the establishment of the right standards and to offer all the aid in our power toward the development of a national spirit in our architecture.   This naturally led to other forms of art, and THE CRAFTSMAN became a magazine for painters and sculptors as well as for architects, interior decorators and craftsmen.   Along these lines it has always been progressive and rather radical, aiming always to discover and bring to the front any notable achievement that seemed to indicate the blazing of a new trail. The magazine has also taken the deepest interest in all social, industrial and political reforms and in the question of industrial education along practical lines that would fit any boy or girl to earn a living under any and all circumstances.   In fact, taken altogether, THE CRAFTSMAN has been the outward and visible expression of the more philosophic side of the Craftsman idea, just as the houses and their furnishings have put into form its more concrete phases.

We have also paid a good deal of attention to agriculture in THE CRAFTSMAN, taking it up along very general lines.   But we have felt that this is a field which required much more exhaustive treatment than we are able to give it in the pages of a magazine of this character.   So to meet this need, we are about to establish a second magazine that will be called "THE YEOMAN" and that will be devoted entirely to the interests of farming, the possibilities of life in rural communities and to the handicrafts that might profitably be carried on in connection with agriculture.

THE time has come, however, to test out all the principles we have been advocating and give them the most practical and comprehensive demonstration within our power.   Therefore we are this year opening a country place, where everything we have said can be put to the test of practical experience.

THE CRAFTSMAN IDEA

This place is called "Craftsman Farms" and it serves as a most complete exposition of the Craftsman idea as a whole. "Craftsman Farms" is situated in the hill country of New Jersey, and our intention is to make it a summer home and school for students of farming and handicrafts. While grown people are welcome, the chief object of the school's existence is to provide an opportunity for the instruction of boys and girls whose parents desire for them a method of training that will enable them to earn a living in whatever circumstances they may happen to be placed. In other words, we purpose to teach them to work with their hands,—not to hoe, dig, plow, or chop wood in a mechanical way,—but to do the kind of constructive work which requires direct thought and which will train them to cope with all the practical problems of life. In fact, the plan of the school is not only a return to the old system of apprenticeship, where the student learned his trade by mastering its difficulties one by one under the guidance of a master craftsman, but it is apprenticeship on improved lines because, instead of working for the benefit of the master, the student acquires by working solely for the sake of his own thorough training and the development in himself not merely skill but initiative and self-reliance.

The instruction will be in the form of lectures and informal talks from the teachers, who will not only give in this way such theoretical knowledge as seems to be required but will answer all questions and respond to all suggestions, so that the student's brain is necessarily made alert by his being forced to take an active part in his own training. The method of instruction will be the same throughout, whether the subject be agriculture, landscape gardening, house building, designing, or any one of the handicrafts. In the latter, students will work side by side with experienced craftsmen, so that every lesson will be the solving of some problem and the doing of actual work according to the methods employed by the best workmen.

For example, every man, sooner or later, hopes or intends to build himself a home. Imagine what that home might be if, as a boy, he had been trained to have a practical working knowledge of drawing, of construction, of the quality and use of different woods, of finishing these woods so that their full value would be brought out and of laying out the grounds surrounding his house so that the most harmonious environment would be a matter of course. As the thing stands now, most men hire some one else to do this sort of thing, which practically amounts to hiring some one else to think for them in matters that most intimately concern their personal life and surroundings.

It is now being very generally acknowledged that our present methods of education fail to a serious degree in the vital work of educating boys and girls toward the larger business of life,—toward the understanding of how to do, and the ability to do, those things upon which our physical existence depends. It does not by any means follow that training along lines of practical work will confine the future activities of the boy to manual labor or to the necessity of doing things for himself. He may use his abilities in as many other directions as he can, but we believe that learning to do the actual work of daily life in the country gives him a kind of ability that may be applied to any form of work, mental or manual, with the best effect, and also that any one possessing it may at any time get back to first principles and start afresh. It has always seemed to us that the great disintegrating force in our modern way of living lies in the system by which everything is done by rote,—and largely by machinery,—and where labor of all kinds is so specially divided that a man, whether he be workman or director, has very little chance to cope with problems outside of his own particular line of work. The great purpose in life and work is the development of character and it naturally follows that true development can come only by the training and use of all the faculties in coping with all the problems that may come up in the ordinary course of life.

In addition to the instruction given at "Craftsman Farms," the conditions of life there will be such as to carry out the same idea. The students, whether young or old, will be housed in small hamlets scattered about the neighborhood in places chosen on account of their fitness for the several things to be done. Each group of cottages will be under the care of a house mother and an instructor and the student will go from hamlet to hamlet until, at the end of the course, he is not only master of the trade he has chosen to learn but also has a general knowledge of related trades and of farming. For example, most of the cottages in which the students live will be designed and built with their active assistance, as the students will be invited to use their own brains and creative ability in designing houses and cottages that they would like to live in or that seem suited to the place. In doing this, of course, they will work directly with the corps of architects that has in charge the designing of the cottages, and in the actual building the students will be allowed to work side by side with experienced carpenters, stone masons, wood finishers, cabinetmakers, blacksmiths and coppersmiths, so that every lesson will be the doing of actual work in the way it is done by competent workmen.

Aside from the educational feature of this enterprise, one of the main objects in carrying it out along the lines indicated is to put to a practical test our favorite theory of a farming community grouped around a central settlement where all social interchange and recreation are as full and convenient as they would be in the city and where every house is within easy reach of the farm lands that belong to it.

If the experiment should prove a success, we confidently look to see it put into practice by many other people; and if it should not, at least we shall have discovered its weak points and have learned something by experience. In any case, the school is meant to complete the work begun by the magazine and to give to the world the result of all the experience we have gained since the first inception of the Craftsman idea ten years ago. If it should have ever so little influence in bringing about the development of our national life along the lines laid down by the men who founded the Republic, it will have fulfilled its mission, because a truth which one man finds courage to utter today is echoed and applied by thousands tomorrow.